Food Science and Technology

Food Science and Technology

Bioprospecting of Natural Compounds in Food, Pharmaceutical and Biomedical Science
Siddhartha Pati, PhD (Editor)
Tanmay Sarkar, PhD (Editor)
Dibyait Lahiri, PhD (Editor)
2022. ISBN: 979-8-88697-361-7 (Hardcover)
2022. ISBN: 979-8-88697-431-7 (eBook)

Proceedings of BIOSPECTRUM: The International Conference on Biotechnology and Biological Sciences: Biotechnological Intervention Towards Enhancing Food Value
Sanket Joshi, PhD (Editor)
Susmita Mukherjee (Editor)
Moupriya Nag, PhD (Editor)
2022. ISBN: 978-1-68507-985-7 (Hardcover)
2022. ISBN: 979-8-88697-069-2 (eBook)

Nutraceuticals: Food Applications and Health Benefits
Anita Kumari, PhD (Editor)
Gulab Singh, PhD (Editor)
2022. ISBN: 978-1-68507-488-3 (Hardcover)
2022. ISBN: 978-1-68507-512-5 (eBook)

A Closer Look at Polyphenolics
Peter Bertollini (Editor)
2022. ISBN: 978-1-68507-480-7 (eBook)
2021. ISBN: 978-1-68507-434-0 (Hardcover)

Food Processing: Advances in Research and Applications
Myriam Huijs (Editor)
2022. ISBN: 978-1-68507-570-5 (Hardcover)
2022. ISBN: 978-1-68507-581-1 (eBook)

More information about this series can be found at
https://novapublishers.com/product-category/series/food-science-and-technology-series/

Gulcin Yildiz
and
Gökçen Yıldız
Editors

Power Ultrasound and Its Applications in Food Processing

Copyright © 2023 by Nova Science Publishers, Inc.
https://doi.org/10.52305/NOGP9909

All rights reserved. No part of this book may be reproduced, stored in a retrieval system or transmitted in any form or by any means: electronic, electrostatic, magnetic, tape, mechanical photocopying, recording or otherwise without the written permission of the Publisher.

We have partnered with Copyright Clearance Center to make it easy for you to obtain permissions to reuse content from this publication. Please visit copyright.com and search by Title, ISBN, or ISSN.

For further questions about using the service on copyright.com, please contact:

Copyright Clearance Center
Phone: +1-(978) 750-8400   Fax: +1-(978) 750-4470   E-mail: info@copyright.com

NOTICE TO THE READER

The Publisher has taken reasonable care in the preparation of this book but makes no expressed or implied warranty of any kind and assumes no responsibility for any errors or omissions. No liability is assumed for incidental or consequential damages in connection with or arising out of information contained in this book. The Publisher shall not be liable for any special, consequential, or exemplary damages resulting, in whole or in part, from the readers' use of, or reliance upon, this material. Any parts of this book based on government reports are so indicated and copyright is claimed for those parts to the extent applicable to compilations of such works.

Independent verification should be sought for any data, advice or recommendations contained in this book. In addition, no responsibility is assumed by the Publisher for any injury and/or damage to persons or property arising from any methods, products, instructions, ideas or otherwise contained in this publication.

This publication is designed to provide accurate and authoritative information with regards to the subject matter covered herein. It is sold with the clear understanding that the Publisher is not engaged in rendering legal or any other professional services. If legal or any other expert assistance is required, the services of a competent person should be sought. FROM A DECLARATION OF PARTICIPANTS JOINTLY ADOPTED BY A COMMITTEE OF THE AMERICAN BAR ASSOCIATION AND A COMMITTEE OF PUBLISHERS.

Library of Congress Cataloging-in-Publication Data

ISBN: 979-8-88697-639-7

Published by Nova Science Publishers, Inc. † New York

*To our lovely dad, İsmail Yıldız,
for his love and support*

Contents

Preface		ix
Chapter 1	**Applications of Ultrasound in Refining Vegetable Oil**	1
	Gökçen Yıldız	
Chapter 2	**Ultrasound Pre-Treatment in Food Drying**	17
	Oguzhan Barel and Gulcin Yildiz	
Chapter 3	**The Effect of Sonication Treatment on the Quality Properties of Apple-Carrot Juice**	37
	Pınar Altaş and Gökçen Yıldız	
Chapter 4	**Ultrasound-Assisted Extraction of Pectin from Tropical Fruit By-Products**	63
	Leticia Xochitl López-Martínez, Manuel Alejandro Vargas-Ortiz and Ebber Addí Quintana-Obregón	
Chapter 5	**Value Added Ultrasound Technology – Assuring Safety of Seafood**	75
	Vasantha Subramoniam Pramitha and Parameswara Panicker Sreejith	
Chapter 6	**Comparison of Antioxidants and Ultrasound Treatment on the Color, Enzymatic Browning and Bioactive Compounds of Fresh-Cut Radishes**	99
	Arzu Imece	

Chapter 7	**Comparative Analysis of Ultrasonic Probe and Water Bath Systems on the Functional Properties of Whey Protein Isolate** 113	
	Menekse Bulut, Rana Muhammad Aadil and Gulcin Yildiz	

Index 129

About the Editors 131

Preface

This book includes seven chapters about power ultrasound and its applications in food processing. Chapter One describes the application of ultrasound technology in refining vegetable oils. Chapter Two provides data related to the effect of ultrasound pre-treatment in drying food. Chapter Three presents the effect of sonication treatment on the quality properties of apple-carrot juice. Chapter Four summarizes the ultrasound-assisted extraction of pectin from tropical fruit by-products. Chapter Five discusses the value-added ultrasound technology assuring safety of seafood. Chapter Six details the comparison of antioxidants and ultrasound treatment on the color, enzymatic browning and bioactive compounds of fresh-cut radishes. Chapter Seven highlights the comparative analysis of ultrasonic probe and water bath systems on the functional properties of whey protein isolate.

Chapter 1

Applications of Ultrasound in Refining Vegetable Oil

Gökçen Yıldız[*]
Department of Food Engineering, Bursa Technical University, Bursa, Turkey

Abstract

Ultrasound applied in the refining stages provides most of the activation energy required for the reaction to occur due to the cavitation force. In the light of the theoretical and practical information obtained in the studies, ultrasound application has been evaluated as an innovative technique due to the shortening of the processing time, low energy consumption, reducing the use of chemicals, and minimal effect on the bioactive components in the oil's structure. Therefore, it can be concluded that ultrasound application can provide significant advantages in the removal of sticky substances from the vegetable oil refining stages and in the deacidification stages as well as in the bleaching stages. As a result, it is thought that it is necessary to accelerate new scientific studies on the use of ultrasound applications in vegetable oil refining stages.

In this review, the principles of ultrasound application, ultrasound application methods, removal of sticky substances in the refining of vegetable oils, ultrasound applications in degumming, deacidification and bleaching stages, and scientific studies on this subject were examined in detail.

Keywords: bleaching, deacidification, degumming, ultrasound, vegetable oils

---

[*] Corresponding Author's Email: gokcen.yildiz@btu.edu.tr.

In: Power Ultrasound and Its Applications in Food Processing
Editors: Gulcin Yildiz and Gökçen Yıldız
ISBN: 979-8-88697-639-7
© 2023 Nova Science Publishers, Inc.

Introduction

Increasing environmental pollution and climate changes in recent years have adversely affected various areas of society. In the food industry, from production to consumption, for the purpose of production of qualified food products without destroying the nature, the search for environmentally friendly practices has begun in stages such as procurement, processing and distribution. As a result of these searches, the concept of "green technology," which can be used for many purposes in the vegetable oil industry, has emerged [1]. With the development of innovative green technologies, the amount of waste generated especially in the refining stages of oils is reduced, while the loss of bioactive components in the oil's structure can be minimized.

Today, many researches are carried out and new technologies are developed in order to reduce the processing time in food production stages, to save energy, and to increase the shelf life and quality of food. Some technological applications (microwave heating, vacuum cooling technology, high pressure and pulsed electric field applications, etc.) have been limited to pilot-scale applications due to high investment costs, consumer acceptance and some legal obstacles.

The ease of accessibility and low investment cost of ultrasound equipment has increased the interest in the use of ultrasound in food production flow [2-6]. Ultrasound is used widely in shortening the processing time, accelerating heat transfer in thermal processing, increasing the rate of chemical reaction, cooking vegetables and meat in order to improve the sensory quality of foods [7-11], improving extraction efficiency [12], drying fruits and vegetables [13-15], filtration of various extracts and beverages [16, 17], in the production of emulsified products such as mayonnaise and ketchup [18] in the defoaming process in fermented beverages [19] and in the separation of milk fat [20] to increase emulsion stability. In addition, scientific studies are carried out on ultrasound application in the extraction of seed and fruit oils, the refining of vegetable oils, especially in the removal of sticky substances, deacidification and decolorization, enzymatic interesterification and emulsification processes, which are oil modification methods.

In this review, the principles of ultrasound application, ultrasound application methods, removal of sticky substances in the refining of vegetable oils, ultrasound applications in degumming, deacidification and bleaching stages, and scientific studies on this subject were examined in detail.

Principle and Application Methods of Ultrasound

Ultrasound, which is among the most used applications within the scope of green technology, is an energy consisting of mechanical vibrations whose frequency is above the hearing thresholds of humans, and they are usually sounds with a frequency greater than 20 kHz [21-23]. Ultrasound, which is a sound wave, propagates in the wave direction in the form of longitudinal waves in liquids and gases. These pressure waves formed by sonic vibrations create acoustic pressure in a time and frequency dependent manner [24, 25]. If this pressure is equal to the negative pressure formed in the relaxation cycle of the wave, cavities in the form of micro bubbles occur in the liquid. The formation of microbubbles in the liquid with the effect of ultrasound is called as "cavitation." The bubbles can grow and expand depending on the intensity of the applied ultrasound waves. These bubbles constantly collide with each other, creating a temporary hot spot with elevated local temperature and pressure estimated at 5000 K and 5000 atm, thereby significantly accelerating chemical reactivity. Thanks to this strong local energy, the chemical activation energy required for the realization of chemical reactions can be provided [26-30].

Ultrasound acts in different frequency ranges. Use of ultrasound in the form of high frequency (100 kHz – 1 MHz) and low density (usually <1 W cm^{-2}) applications that do not adversely affect the quality characteristics of foods. Low-density treatments are generally used as an analytical technique to obtain information about the physicochemical properties of foods such as hardness, maturity, sugar content and acidity [31]. Low-frequency processes are applied to physically or chemically change food properties [32]. Ultrasound applications with a frequency of 20–40 kHz and an applied power of >10 W cm^{-2} is used in the preparation of oil emulsions (40-200 nm) and extraction of various bioactive and organic compounds [33]. To determine the effectiveness of the ultrasound process, parameters such as frequency, wavelength, amplitude and ultrasonic power need to be optimized [34]. In addition, the reactor design and the method of application of ultrasound also determine the efficiency of the process [35].

Ultrasound is applied with two different methods, namely ultrasonic bath and ultrasonic probe. Ultrasonic baths are generally used to remove air from solvents or to clean small glassware. Ultrasonic baths are rarely used in food processing streams because the reproducibility of chemical reactions is very low and the energy density transferred is low [36]. Another method is the application of ultrasound with ultrasonic probes. In this method, ultrasonic intensity is transferred to a smaller surface (only the probe tip) as the probe is immersed in the sample. This method is more effective than ultrasonic baths and is used in small volume production [2].

Ultrasound Applications in the Refining of Raw Vegetable Oils

Vegetable seed and fruit oils other than natural olive oil cannot be consumed without refining, as they contain undesirable components (free fatty acids, various color and fragrance substances, aldehydes and ketones, hydrocarbons, etc.) in their structure. The refining process consists of the steps applied in order to remove the unwanted compounds in the structure of the raw vegetable oil or to reduce their amounts to an acceptable level [37]. Chemical or physical methods are used in the refining of raw oils. Chemical refining is consisting of the stages as removal of sticky substances (degumming), deacidification, decolorization and deodorization [38]. In the physical refining process, there are degumming, decolorization stages and water vapor distillation stages where deacidification and deodorization processes are carried out simultaneously [39].

During the refining process, the unwanted compounds in the raw oil are removed. Depending on the processing conditions, negative changes occur in the structure of the oil, and the bioactive components in the oil suffer qualitative and quantitative losses [40]. Many studies have been conducted on the loss of bioactive substances during the refining of vegetable oils [41-43]. Bioactive components are substances with antioxidant and functional properties. Therefore, it is very important to establish innovative refining strategies that will prevent the bioactive components from leaving the oil in the refining stages [44]. Ultrasound application is among the new approaches applied to minimize the loss of bioactive components in raw oil during refining.

In below, the results of scientific studies on ultrasound applications in the removal of sticky substances-degumming, deacidification and bleaching stages of chemical refining were discussed in details.

Removal of Sticky Substances (Degumming)

Raw oil contains phospholipids, proteinaceous compounds, gums and resin-like adhesives. Phospholipids contain a glycerol molecule, two fatty acids, and a negatively charged phosphate group modified by an alcohol group. The polar ends of the phospholipids containing the phosphate group are hydrophilic (water-loving), while the fatty acid chains are apolar and hydrophobic. Phospholipids exist in two forms, hydratable and non-hydratable [45]. Since sticky substances cause problems in alkaline neutralization, bleaching and deodorization stages, the sticky substances that constitute the first stage of the refining process are removed from the oil by the degumming process.

According to the structural properties of phospholipids (hydratable or non-hydratable), different methods are applied in the degumming process. These methods are:

- The hydration method which is the removal of sticky substances with low pH solutions (phosphoric acid, acetic anhydride, nitric acid, etc.), and
- the enzymatic degumming method that has been used in recent years [46-48].

Another alternative method is membrane degumming. Membrane technology has many advantages over traditional separation methods. These advantages can be listed as:

- the separation of phospholipids with the membrane at room temperature, thus obtaining higher quality products for temperature-sensitive products,
- application in combination with other processes,
- being suitable for continuous or batch systems,
- not using chemicals, and
- low energy consumption [49-50].

The hydratable phospholipids are easily removed by the hydration method. On the other hand, non-hydratable phospholipids lose their lipophilic properties when low pH solutions are used and turn into a hydrated form and are removed from the oil by acid degumming-hydration method [45].

Another method applied for the removal of phospholipids is the enzymatic degumming method [51]. Especially in recent years, developments in biotechnological studies have led to an increase in studies related to the use of enzymes in the oil industry. In the degumming process, enzymatic degumming method is also applied to reduce the amount of phospholipid in the raw oil to the critical value (≤10 ppm) [52]. Advantages of the enzymatic degumming process over the traditional method is stated as a decrease in the amount of alkali used during alkali neutralization, an increase in efficiency, and a decrease in the amount of waste water [53]. However, the long processing time in enzymatic methods is the most crucial issue. For this reason, scientific studies have been carried out to determine the effect of the use of ultrasound in the enzymatic degumming method on enzyme activity and reaction rate.

More and Gogate [54] examined the effects of different parameters (enzyme amount, pH, water amount, temperature and ultrasound power) on the ultrasound assisted-enzymatic degumming process and explained the effectiveness of the degumming process with the "extent of degumming." As a result of the evaluation made for each parameter, the optimum conditions for enzymatic degumming with ultrasound assistance were determined as:

- 2.0 ml/L enzyme amount (phospholipase A enzyme),
- pH: 5,
- water amount: 5%,
- processing time: 120 minutes, and
- ultrasound power: 100 W. Under optimum conditions, extent of degumming was calculated as 98.38%. In addition, it was determined that ultrasound application decreased the free fatty acidity and totox value of the oil in terms of oil quality.

Ultrasound assisted-acid degumming process (phosphoric acid) was applied in raw soybean oil and sunflower oil in the study of Mahmood-Fashandi et al. [55]. It was determined that the processing time was shortened significantly in the ultrasonic assisted-acid degumming process performed with an ultrasonic bath. They stated that the ultrasound application did not create a statistically significant difference in the fatty acid compositions of the oils, and the change in free fatty acid and oxidation values did not cause a problem and could be eliminated in other stages of refining.

Jiang et al. [56] investigated the effect of the combination of mechanical mixing and mechanical mixing–ultrasound application on the processing conditions in the enzymatic degumming process with rapeseed oil in an ultrasonic bath. It was stated that the required amount of water and enzyme (phospholipase A1) decreased, the process was carried out effectively in a shorter time, and there was no statistically significant difference between the two applications in terms of free fatty acidity and peroxide value of the oil when mechanical mixing-ultrasound combination is applied.

Liu et al. [57] stated in their study that ultrasound application increased enzyme activity and reaction rate with the impact of cavitation.

Deacidification

Alkaline solutions such as sodium hydroxide, potassium hydroxide, sodium carbonate is used commonly in the deacidification process with alkalis [58]. Free fatty acids enter into saponification reaction with alkali solution and soap phase is formed. The soap phase formed gives the oil/water emulsion a more durable structure [59]. In particular, the loss of neutral oil increases due to the use of more alkali than the amount of alkali determined according to the free fatty acid content of the oil. It was aimed to deacidify raw soybean oil by using different concentrations of alkali solution attached to an ultrasonic water bath and adsorbents [60-62]. Kieselguhr and silica were used as adsorbent and sodium hydroxide as alkali. In the ultrasound assisted deacidification process, the free acidity value of soybean oil and the specific absorption values in UV light were decreased. In the ultrasound assisted deacidification process, the neutral oil yield was determined as 86.57% and the oil was lightened in color. The results showed that there is no need to use more than 10% of the theoretically calculated alkaline amount and energy saving is achieved since ultrasound application is used for heating the oil.

In order for the saponification reaction between the alkali solution and free fatty acids to occur effectively, the surface area where the oil and alkali come into contact must increase. It is thought that ultrasound application may have a positive effect in the deacidification stage in terms of providing an effective mixing. However, there has not been enough scientific study on the application of ultrasound assisted deacidification.

Bleaching

The bleaching process, which is one of the most important stages of the refining process, is the process of removing color substances and some oxidation products from the oil with the help of an adsorbent [63, 64]. In addition, phospholipids, trace metals and soap residues remaining in the oil are also removed from the oil during the bleaching process [65]. Adsorbents such as natural bleaching earths, acid activated bleaching earths, activated carbon and amorphous silicate are used for the bleaching process [66]. The bleaching process is carried out at 85-100°C and under vacuum of 50-100 mmHg [67].

In recent years, studies on ultrasound assisted bleaching have increased. Liang et al. [68] investigated the effect of ultrasound assisted-bleaching on the chlorophyll and carotene content of hemp seed oil and the change in oxidative stability. For this purpose, three different adsorbents (sepiolite, active bentonite and industrial bleaching earth) were used and ultrasound assisted-bleaching was applied with an ultrasonic probe. At the end of the treatment, it was observed that there was a significant decrease in the chlorophyll and carotene content. It was observed that the highest reduction was achieved with active bentonite soil, and it was found that significant reduction in peroxide number and conjugated diene values were achieved in all adsorbents. It was observed that the formation of primary oxidation products in the samples that were bleached by ultrasound assisted-bleaching application was slower than the control samples.

In the study by Icyer and Durak [69] about ultrasound assisted- and conventional bleaching processes in canola oil, the oil was compared in terms of color values and quality criteria. In the study where the ultrasonic probe was used, the desired color values were achieved in 5% less time. Ultrasound assisted-bleaching was applied at a lower temperature (25% lower) than the traditional method. It was observed that there was no difference between the two methods in terms of oxidative reactions. Optimum trading conditions were determined as:

- 75°C temperature,
- 60% ultrasound power,
- 0.8 g (mass/mass) soil content, and
- 15 minutes treatment time.

Abbasi et al. [70] applied conventional and ultrasound assisted-bleaching processes in sunflower and olive oils in their study. Ultrasound assisted-bleaching process was performed in an ultrasonic bath by applying 2 different temperatures (45 and 60°C) and 3 different times (10, 20 and 30 minutes). Conventional bleaching was carried out at 110°C for 60 minutes. It was observed that the ultrasound assisted-bleaching process was more effective in reducing the chlorophyll and carotene content of the oil compared to the traditional method. Ultrasound application did not change the fatty acid compositions of the oils. It was stated that the process time is shortened, the process temperature is reduced, energy consumption is reduced and the bleaching process is realized effectively with the use of ultrasound.

Abedi et al. [71] investigated the effect of ultrasound assisted-bleaching on the chemical and physical properties of soybean oil. They stated that ultrasound treatment at a maximum power of 400 W and a frequency of 25 kHz, at a power level of 45-90%, accelerated the breakdown of chlorophylls and carotenoids in soybean oil. Due to this effect, it has been stated that the desired color values can be achieved by using less amount of bleaching earth in the ultrasound assisted-bleaching process. It was emphasized that performing the bleaching process with ultrasound assistance, especially in terms of less bleaching earth waste, can have an important place in oil refining.

Hosseini et al. [72] investigated the effects of ultrasound assisted-bleaching applied to olive oil, sunflower oil and sesame oil on some physical and chemical properties of the oil with a 400 W ultrasound device at 25, 60 and 100% power. The results of the study showed that as the power level increases in ultrasound application, a more effective color changes (generally a decrease in L^* value) is achieved. In addition, a significant decrease was achieved in the chlorophyll and β-carotene contents of the oils in parallel with the increase in ultrasound power. Ultrasound application did not change the viscosity and density of the oil samples. In addition, it was determined that ultrasound treatment caused a minimal increase in the peroxide number of the oil, while it did not increase the free fatty acidity.

In a similar study with ultrasound assistance, researchers have applied bleaching process in rapeseed oil using different bleaching earths and without using bleaching earth in an ultrasound bath operating at 20 kHz frequency [73]. In this study, color change in rapeseed oil after ultrasound assisted-bleaching was investigated by determining the visible region color spectrum obtained by spectrophotometer. According to the results of the research, it was determined that the color reduction of rapeseed oil was achieved at the same level as the conventional bleaching process performed with the addition of bentonite at 130°C, as a result of ultrasound applied without using an adsorbent at 60% power. This result showed that ultrasound treatment can decompose carotenoids in oils above a certain ultrasound power level. It is stated that the most important effect in the ultrasound assisted-bleaching process is on the lightening time [74].

Conclusion

Ultrasound applied in the refining stages provides most of the activation energy required for the reaction to occur due to the cavitation force. In the light of the theoretical and practical information obtained in the studies, ultrasound application has been evaluated as an innovative technique due to the shortening of the processing time, low energy consumption, reducing the use of chemicals, and minimal effect on the bioactive components in the oil's structure. Therefore, it can be concluded that ultrasound application can provide significant advantages in the removal of sticky substances from the vegetable oil refining stages and in the deacidification stages as well as in the bleaching stages. As a result, it is thought that it is necessary to accelerate new scientific studies on the use of ultrasound applications in vegetable oil refining stages.

References

[1] Güneş E, Keskin B, Kıymaz T. Gıda Sanayiinde Yeşil Ekonomi ve Uygulamaları. XI. *Tarım Ekonomisi Kongre Kitabı/Samsun*. 2014; 1528-1532.

[2] Chemat F, Khan MK. Applications of ultrasound in food technology: processing, preservation and extraction. *Ultrason Sonochem* 2011; 18(4): 813-835.

[3] Lee H, Yildiz G, Dos Santos LC, Jiang S, Andrade J, Engeseth NC, Feng H. Soy protein nano-aggregates with improved functional properties prepared by sequential pH treatment and ultrasonication. *Food Hydrocoll* 2016; 55: 200–209.

[4] Yildiz G, Rababah T, Feng H. Ultrasound-Assisted Cutting of Cheddar, Mozzarella and Swiss Cheeses – Effects on Quality Attributes during Storage. *Innov Food Sci EmergTechnol* 2016; 37: 1-9.

[5] Yildiz G, Andrade J, Engeseth NC, Feng H. Functionalizing soy protein nano-aggregates with pH-shifting and mano-thermo-sonication. *J Colloid Interface Sci* 2017; 505: 836-846.

[6] İzli G, Yildiz G. Evaluation of the high intensity ultrasound pre-treatment effects on the physical properties and bioactive compounds of convective dried quince samples. *Int J Fruit Sci* 2021; 21(1): 645-656.

[7] Yildiz G, Ding J, Andrade J, Engeseth NJ, Feng H. Effect of plant protein-polysaccharide complexes produced by mano-thermo-sonication and pH-shifting on the structure and stability of oil-in-water emulsions. *Innov Food Sci Emerg Technol* 2018; 47: 317-325.

[8] Zou Y, Kang D, Liu R, Qi J, Zhou G, Zhang, W. Effects of ultrasonic assisted cooking on the chemical profiles of taste and flavor of spiced beef. *Ultrason Sonochem* 2018; 46: 36-45.

[9] Yildiz G. Application of ultrasound and high-pressure homogenization against high temperature-short time in peach juice. *J Food Process* Eng 2019; 42(3): e12997.

[10] Yildiz G, Palma S, Feng H. Ultrasonic cutting as a new method to produce fresh-cut red delicious and golden delicious apples. *J Food Sci* 2019; 84(12): 3391-3398.

[11] Yildiz G, Feng H. Sonication of Cherry Juice: Comparison of Different Sonication Times on Color, Antioxidant Activity, Total Phenolic and Ascorbic Acid Content. *Lat Am Appl Res* 2019; 49(4),:255-260.

[12] Khoei M, Chekin F. The ultrasound-assisted aqueous extraction of rice bran oil. *Food Chem* 2015; 194: 503–507.

[13] Tao Y, Zhang J, Jiang S, Xu Y, Show PL, Han Y, Ye M. Contacting ultrasound enhanced hot-air convective drying of garlic slices: Mass transfer modeling and quality evaluation. *J Food Eng* 2018; 235: 79-88.

[14] Yildiz G, Izli G. The effect of ultrasound pretreatment on quality attributes of freeze-dried quince slices: Physical properties and bioactive compounds. *J Food Process Eng* 2019; 42 (5): e13223.

[15] Yıldız G, İzli G, Çavuş M, Ceylan MM. The Effect of Ultrasound Pre-treatment on the Quality Charcteristics of Dried Iğdır Apricot. *Journal of the Institute of Science and Technology*, 2021; 11(1): 303-313.

[16] Liu D, Vorobiev E, Savoire R, Lanoisellé JL. Comparative study of ultrasound-assisted and conventional stirred dead-end microfiltration of grape pomace extracts. *Ultrason Sonochem* 2013; 20(2): 708-714.

[17] Yildiz G, Aadil RM. Comparison of high temperature-short time and sonication on selected parameters of strawberry juice during room temperature storage. *J Food Sci Technol* 2020; 57(4): 1462-1468.

[18] Gavahian M, Chen YM, Khaneghah AM, Barba FJ, Yang BB. In-pack sonication technique for edible emulsions: Understanding the impact of acacia gum and lecithin emulsifiers and ultrasound homogenization on salad dressing emulsions stability. *Food Hydrocoll* 2018; 83: 79-87.

[19] Mawson R, Tongaonkar J, Bhagwat SS, Pandit AB. Airborne ultrasound for enhanced defoaming applications. *Innov Food Process Technol* 2016; 347-359.

[20] Johansson L, Singh T, Leong T, Mawson R, McArthur S, Manasseh R, Juliano P. Cavitation and non-cavitation regime for large-scale ultrasonic standing wave particle separation systems–In situ gentle cavitation threshold determination and free radical related oxidation. *Ultrason Sonochem* 2016; 28: 346-356.

[21] Duran K. Bahtiyari Mİ, Körlü A. E, Dereli S, Özdemir D. Ultrasound technology. *Textile and Apparel* 2006; 16(3): 155-158.

[22] Dinçer C, Topuz A. The use of ultrasound in fruit juice processing. *The Journal of Food*, 2018; 43(4): 569-581.

[23] Yildiz G, Izli G, Aadil RM. Comparison of chemical, physical, and ultrasound treatments on the shelf life of fresh-cut quince fruit (Cydonia oblonga Mill.). *J Food Process Preserv* 2020; 44 (3): e14366.

[24] Tüfekçi S, Özkal SG. Effect of Ultrasound Pre-Treatment on Drying Characteristics of Carrot Slices. *The Journal of Food* 2018; 16(1): 11-19.

[25] Yildiz G. The Effect of High Intensity Ultrasound Pre-treatment on the Functional Properties of Microwave-dried Pears (*Pyrus communis*). *Lat Am Appl Res* 2021; 51(2): 133-137.
[26] Ince NH, Tezcanli G, Belen RK, Apikyan IG. Ultrasound as a catalyzer of aqueous reaction systems: the state of the art and environmental applications. *Appl Catal B: Environmental* 2001; 29(3): 167-176.
[27] Pingret D, Fabiano-Tixier AS, Chemat F. Degradation during application of ultrasound in food processing. *Food Control* 2013; 31(2): 593-606.
[28] Kuleaşan Ş, Şahin K. Ecologic Approaches in Edible Oil Refining: Applications of Ultrasound Technology. *Academic Food Journal* 2015; 13(2):149-155.
[29] Yıldız G. Physicochemical properties of soy protein concentrate treated with ultrasound at various amplitudes. *Journal of the Institute of Science and Technology*, 2018; 8(4): 133-139.
[30] Jiang S, Yildiz G, Ding J, Andrade J, Rababah TM, Almajwalc A, Abulmeatyc MM, Feng H. Pea Protein Nanoemulsion and Nanocomplex as Carriers for Protection of Cholecalciferol (Vitamin D3). *Food Bioproc Technol* 2019; 12(6): 1031-1040.
[31] Demirdoven A, Baysal T. The use of ultrasound and combined technologies in food preservation. *Food Rev Int* 2009; 25(1): 1-11.
[32] Soria AC, Villamiel M. Effect of ultrasound on the technological properties and bioactivity of food: a review. *Trends Food Sci Technol* 2010; 21(7): 323-331.
[33] Ashokkumar M. Applications of ultrasound in food and bioprocessing. *Ultrason Sonochem* 2015; 25: 17-23.
[34] Yildiz G, Aadil RM. Comparative Analysis of Antibrowning Agents, Hot Water and High-intensity Ultrasound Treatments to Maintain the Quality of Fresh-cut Mangoes. *J Food Sci Technol* 2022; 59(1): 202-211.
[35] Esclapez MD, García-Pérez JV, Mulet A, Cárcel JA. Ultrasound- assisted extraction of natural products. *Food Eng Rev* 2011; 3(2): 108-120.
[36] Wen C, Zhang J, Zhang H, Dzah CS, Zandile M, Duan Y, Luo X. Advances in ultrasound assisted extraction of bioactive compounds from cash crops-A review. *Ultrason Sonochem* 2018; 48: 538-549.
[37] Yemişçioğlu F, Özdikicierler O, Gümüşkesen AS, Sönmez AE. Re-engineering of edible oil refining wastes. *The Journal of Food* 2013; 38(6): 367-374.
[38] Chew SC, Tan CP, Long K, Nyam KL. Effect of chemical refining on the quality of kenaf (Hibiscus cannabinus) seed oil. *Ind Crops Prod* 2016; 89: 59-65.
[39] Gümüşkesen AS, Yemişçioğlu F. Bitkisel Sıvı ve Yağ Üretim Teknolojisi. *Meta Basım Matbaacılık Hizmetleri*. ISBN: 975-94208-0-5, İzmir, 2010; 216 s.
[40] El-Mallah MH, El-Shami SM, Hassanien MMM, Abdel-Razek AG. Effect of chemical refining steps on the minor and major components of cottonseed oil. *Agric Biol JN Am* 2011; 2: 341-349.
[41] Ghazani SM, Marangoni AG. Minor components in canola oil and effects of refining on these constituents: A review. *J Am Oil Chem Soc* 2013; 90(7): 923-932.
[42] Ghazani SM, Garcia-Llatas G, Marangoni AG. Minor constituents in canola oil processed by traditional and minimal refining methods. *J Am Oil Chem Soc*, 2013; 90(5): 743-756.

[43] Kreps F, Vrbiková L, Schmidt Š. Influence of industrial physical refining on tocopherol, chlorophyll and beta-carotene content in sunflower and rapeseed oil. *Eur J Lipid Sci Technol* 2014; 116(11): 1572-1582.
[44] Ergönül PG, Köseoğlu O. Changes in α-, β-, γ-and δ-tocopherol contents of mostly consumed vegetable oils during refining process. *CyTA-J Food*, 2014; 12(2): 199-202.
[45] Paisan S, Chetpattananondh P, Chongkhong S. Assessment of water degumming and acid degumming of mixed algal oil. *J Environ Chem Eng* 2017; 5(5): 5115-5123.
[46] Sengar G, Kaushal P, Sharma HK, Kaur M. Degumming of rice bran oil. *Rev Chem Eng* 2014; 30(2): 183-198.
[47] Chew SC, Tan CP, Nyam KL. Optimization of degumming parameters in chemical refining process to reduce phosphorus contents in kenaf seed oil. *Sep Purif Technol* 2017; 188: 379-385.
[48] Vlahopoulou G, Petretto GL, Garroni S, Piga C, Mannu A. Variation of density and flash point in acid degummed waste cooking oil. *J Food Process Preserv* 2018; 42(3): e13533.
[49] Firman LR, Ochoa NA, Marchese J, Pagliero C. (2013). "Deacidification and solvent recovery of soybean oil by nanofiltration membranes. *J Membr Sci* 2013; 431: 187-196.
[50] Boynueğri P, Yemişçioğlu F, Gümüşkesen AS. Effect of membrane degumming conditions on permeate flux and phospholipids rejection. *The Journal of Food* 2017; 42(5):597-602.
[51] Sampaio KA, Zyaykina N, Wozniak B, Tsukamoto J, Greyt WD, Stevens CV. Enzymatic degumming: degumming efficiency versus yield increase. *Eur J Lipid Sci Technol* 2015; 117(1): 81-86.
[52] Turetkan G, Yucedag CT, Ustun G, Tuter M. Enzymatic degumming process for crude corn oil with phospholipase A1. *Int J Eng Sci* (IJSES) 2018; (4): 1-14.
[53] Mei L, Wang L, Li Q, Yu J, Xu X. Comparison of acid degumming and enzymatic degumming process for Silybum marianum seed oil. *J Sci Food Agric* 2013; 93(11): 2822-2828.
[54] More NS, Gogate PR. Ultrasound assisted enzymatic degumming of crude soybean oil. *Ultrason Sonochem* 2018; 42, 805-813.
[55] Mahmood-Fashandi H, Ghavami M, Gharachorloo M, Abbasi R, Mousavi Khaneghah A. Using of Ultrasonic in Degumming of Soybean and Sunflower Seed Oils: Comparison with the Conventional Degumming. *J Food Process Preserv* 2017; 41(1): e12799.
[56] Jiang X, Chang M, Wang X, Jin Q, Wang X. The effect of ultrasound on enzymatic degumming process of rapeseed oil by the use of phospholipase A1. *Ultrason Sonochem* 2014; 21(1): 142-148.
[57] Liu Y, Jin Q, Shan L, Liu Y, Shen W, Wang X. The effect of ultrasound on lipase-catalyzed hydrolysis of soy oil in solvent-free system. *Ultrason Sonochem* 2008; 15(4): 402-407.
[58] Engelmann JI, Ramos LP, Crexi VT, Morais MM. Degumming and neutralization of rice bran oi. *J Food Process Eng* 2017; 40(2): e12362.

[59] Yooritphun K, Lilitchan S, Aryusuk K, Krisnangkura K. Effect of Selected Polyhydric Alcohols on Refining Oil Loss in the Neutralization Step. *J Am Oil Chem Soc* 2017; 94(2): 301-308.

[60] Kuleaşan Ş, Demirok H, Çınar A. Ham Soya Yağının Ultrases Eşliğinde Asitliğinin Giderilmesi. YABİTED III. *Bitkisel Yağ Kongresi*, 2017; s.58, 13-15 Nisan, İzmir-Hilton.

[61] Kuleaşan Ş, Demirok H, Güler HÖ, Çınar A. Effect of Ultrasound Assisted Alkali Neutralization Process on Fatty Acid Profile of Soybean Oil. *2nd International Tourism and Microbial Food Safety Congress*, Antalya, Turkey 13-14 December 2017.

[62] Gökalp K. *Ham Zeytinyağının Ultrases Etkisi Altında Asitliğinin Giderilmesinin Araştırılması.* Yüksek Lisans Tezi, Mehmet Akif Ersoy Üniversitesi, Burdur Türkiye, 2018.

[63] Strieder MM, Pinheiro CP, Borba VS, Pohndorf RS, Cadaval Jr TR, Pinto LA. Bleaching optimization and winterization step evaluation in the refinement of rice bran oil. *Sep Purif Technol* 2017; 175: 72-78.

[64] Hatami AM, Sabour MR, Amiri A. Recycling process of spent bleaching clay: Optimization by response surface methodology. *Glob J Environ Sci Manag* 2018; 4(1): 9-18.

[65] Chew SC, Tan CP, Nyam KL. Optimization of bleaching parameters in refining process of kenaf seed oil with a central composite design model. *J Food Sci* 2017; 82(7): 1622-1630.

[66] Ismadji S, Soetaredjo FE, Ayucitra A. *Clay materials for environmental remediation.* New York: Springer. 2015; 25: 1-124.

[67] Lee SY, Jung MY, Yoon, SH. Optimization of the refining process of camellia seed oil for edible purposes. *Food Sci Biotechnol* 2014; 23(1): 65-73.

[68] Liang J, Aachary AA, Hydamaka A, Eskin NM, Eck P, Thiyam-Holländer U. Reduction of Chlorophyll in Cold-Pressed Hemp (Cannabis sativa) Seed Oil by Ultrasonic Bleaching and Enhancement of Oxidative Stability. *Eur J Lipid Sci Technol* 2018; 120(4): 1700349.

[69] Icyer N C, Durak M. Ultrasound-assisted bleaching of canola oil; Improve the bleaching process by central composite design. *LWT – Food Sci Technol* 2018; 97: 640-647.

[70] Abbasi R, Gharachorloo M, Ghavami M, Mahmood-Fashandi H, Mousavi Khaneghah A. The effect of ultrasonic waves in bleaching of olive and sunflower oils and comparison with conventional bleaching. *J Food Process Preserv* 2017; 41(4): e13079.

[71] Abedi E, Sahari MA, Barzegar M, Azizi MH. Optimisation of soya bean oil bleaching by ultrasonic processing and investigate the physico-chemical properties of bleached soya bean oil. *Int J Food Sci Technol* 2015; 50(4): 857-863.

[72] Hosseini S, Gharachorloo M, Tarzi BG, Ghavami M, Bakhoda H. Effects of ultrasound amplitude on the physicochemical properties of some edible oils. *J Am Oil Chem Soc* 2015; 92(11-12): 1717-1724.

[73] Su D, Xiao T, Gu D, Cao Y, Jin Y, Zhang W, Wu T. Ultrasonic bleaching of rapeseed oil: Effects of bleaching conditions and underlying mechanisms. *J Food Eng* 2013; 117(1): 8-13.

[74] Jahouach-Rabai W, Trabelsi M, Van Hoed V, Adams A, Verhé R, De Kimpe N, Frikha MH. Influence of bleaching by ultrasound on fatty acids and minor compounds of olive oil. Qualitative and quantitative analysis of volatile compounds (by SPME coupled to GC/MS). *Ultrason Sonochem* 2008; 15(4): 590-597.

Chapter 2

Ultrasound Pre-Treatment in Food Drying

Oguzhan Barel
and Gulcin Yildiz[*]

Department of Food Engineering, Igdir University, Iğdır, Turkey

Abstract

Sound waves that human beings can hear are in the range of 16 Hz to 18 kHz. On the other hand, ultrasonic sound waves are a form of energy that can be produced by sonic waves at or above 20 kHz. It is stated that ultrasound is an acoustic energy source. When an acoustic wave passes in a medium, it causes compression and decompression. Due to this turbulence, a large amount of energy is produced and mass transfer is increased. The main principle is that the sound wave is reflected and scattered similarly to the light wave. Although the maximum limit of ultrasonic sound wave frequency has not been fully proven, it is expressed as 5 MHz for gases and 500 MHz for solids. Ultrasound technology is expressed as power of sound (W), intensity of sound (Wm^{-2}) or energy density of sound (Wsm^{-3}).

Ultrasound technology creates many physical, chemical and biochemical reactions that allow it to be effective in various applications while passing through an environment and occur depending on the frequency and amplitude of sound waves. The cavitation phenomenon, which is the most important one of these effects, is the formation of bubbles as a result of the existing distance between the molecules being above the standards in areas where the sound wave pressure decreases while the cavitation event progresses in a liquid, the volumetric expansion of these bubbles that appear in the materials exposed to

---

[*] Corresponding Author's Email: gulcn86@gmail.com.

In: Power Ultrasound and Its Applications in Food Processing
Editors: Gulcin Yildiz and Gökçen Yıldız
ISBN: 979-8-88697-639-7
© 2023 Nova Science Publishers, Inc.

continuous sound waves, and they will not be able to absorb energy at higher levels. When they reach the maximum size, there is an implosive collapse. In the area where the internal explosions of the cavitation balloons occur, an intense energy produces. This energy causes the environment to heat up and results in chemical reactions. As a result of this, high energy shear waves and turbulences occur in the cavitation area due to the increase in temperature and pressure.

The use of ultrasound technology in the food industry provides positive developments in the drying of food products. In this study, the effect of ultrasonic sound waves, which is one of the green technologies mentioned with its positive contributions to the environment, on the drying of food products is explained with the findings obtained from the literature.

Keywords: cavitation, drying, ultrasonic bath, ultrasound pre-treatment, ultrasonic probe

Introduction

In order to ensure the physical, chemical, biological and microbial safety of food products, thermal processes are frequently applied in the food industry. In the application of thermal processes, as a result of the negative effects of high temperatures and time-based sensory characters and bioactive substances called secondary metabolites, it has been concentrated on novel applications, which are expressed as non-thermal green technologies. These innovative applications that do not show thermal properties include ultrasonic sound waves (ultrasound), pulsed electrical field (PEF), accented light, high pressure (HP) process, micro-filtration, X-ray, ultraviolet light (UV), high voltage oscillating magnetic field.

Ultrasonic sound waves are expressed as sound waves with frequencies higher than what human beings can hear, that is, between 20 kHz and 10 MHz. Ultrasonic sound waves in the food industry is preferred for below reasons:

- less damage to food products compared to thermal applications such as pasteurization and sterilization,
- shortens the time of use of the thermal process,
- accelerating heat transfer in thermal applications,
- increase the rate of chemical reaction,

- In addition to being important in many ways such as improving the sensory quality of food products,
- the equipment is easy to access and cheap in terms of cost

Except for the drying of food products, ultrasound is frequently used for the;

- improvement of the extraction efficiency,
- dehydration of food products,
- filtration of various extracts and beverages,
- improvement of the durability of emulsions,
- removal of foam in fermented beverages,
- coagulation in milk,
- quality analysis of eggs,
- surface cracking of cheese,
- texture of biscuits,
- control of wine fermentation,
- characterization properties of dough, and
- in the analysis of physicochemical factors such as sugar content and acidity.

Ultrasound (Ultrasonic Sound Waves)

The first use of ultrasound was as a result of the collective death of fish by the sound waves sent by submarines during the world wars. During the 1960s, low-frequency and high-energy ultrasonic sound waves became increasingly popular in the industry [1]. The frequency of the sound wave that human beings can hear range between 16 Hz and 18 kHz. On the other hand, ultrasound is expressed as a form of energy that can be produced by sonic waves at the limits of 20 kHz or higher. Ultrasound is a type of energy characterized as acoustic [2]. When acoustic waves propagate in a medium, they cause compression and decompression. Due to this turbulence, a large amount of energy is produced and increases in mass transfer is achieved. The main principle here is the reflection and scattering of sound waves close to the light wave [3]. Although the maximum level of ultrasound frequency has not been fully proven, it is expressed as 5 MHz for gases and 500 MHz for solids.

Ultrasonic sound waves can be defined as:

- the power of the sound (W),
- the intensity of the sound (W/m^2) or
- the energy density of sound (Ws/m^3) [4].

It is possible to see the frequency range of ultrasound in Table 1.

Table 1. Frequency range of ultrasound

Sound waves	Infrasonic	Acoustic	Ultrasound		
Frequency values	2 Hz–20 Hz	16 Hz–20 kHz	20 kHz–20 MHz		
Characteristics	-	16 Hz–18 kHz Human ear	20 kHz–100 kHz Use in food industry	20 kHz–2 MHz Extended range	5 MHz–10 MHz Use in identification

Ultrasonic sound waves are divided inti three groups in terms of frequency:

- power ultrasonic sound waves in the range of 16 kHz and 100 kHz,
- high frequency ultrasonic sound waves between 100 kHz and 1 MHz,
- recognizing ultrasonic sound waves between 1 MHz and 10 MHz

Ultrasonic sound waves are generated either by pressure electricity (piezoelectric) or by magnetic compression transducers that generate high-energy vibrations [5].

Application of ultrasonic sound waves in the food industry occurs in two forms:

- Ultrasound with high intensity and low frequency
- Low-intensity and high-frequency ultrasound

The intensity value of the application, which is called low-intensity ultrasound, is lower than 1 W/m^2 and the frequency value is higher than 100 kHz. This process finds use in many applications stated as below:

- in cleaning surface areas,
- inactivating enzymes,
- in crystallization processes,
- in the emulsification process,
- in the filtration stage,
- in the freezing process and
- in tenderization processes of meat products.

On the other hand, the intensity of high-intensity ultrasonic sound waves is more than 1 W/m^2 and the frequency value is in the range of 18-100 kHz. It is used in processes such as:

- in the deaeration process,
- in oxidation/reduction processes
- in the extraction of enzymes and proteins,
- in processes such as the production of the core structure during both the inactivation and crystallization of enzymes [6].

In addition, high-power ultrasonic sound waves are highly effective in the disintegration of cells, reduction in particle volume, and inactivation of bacterial spores [7].

In the application of low-intensity ultrasonic sound waves, neither physical nor chemical change occurs in the material through which the sound waves pass. The low-energy ultrasound process is most commonly used in determining the physicochemical characteristics of food products such as hardness, maturity, food composition, particle size, and acidity [8].

Ultrasound with high energy is used for microbiological and enzymatic inactivation in food products. While the application of high-energy ultrasonic sound waves affects food products physically, chemically and mechanically, it is not possible to talk about such an effect in the application of low-energy ultrasonic sound waves [9].

While ultrasonic sound waves move in a certain direction, they create many physical, chemical and biochemical reactions that occur depending on the frequency and amplitude values of the sound waves, which allows them to be found in various applications [8]. The cavitation phenomenon (Figure 1), which is at the very beginning of these effects, is the formation of bubbles as a result of the existing distance between the molecules being above the standards in places where the sound wave pressure decreases while passing in a liquid, and the gradual expansion of these bubbles that appear in the material that is constantly exposed to sound waves. It is expressed as oscillation and damping towards the inside when they reach the maximum size that cannot absorb more energy above this value [10, 11]. An intense energetic accumulation occurs in the area where the internal explosion of cavitation bubbles takes place [12]. This accumulated intense energy causes the environment to heat up and results in chemical reactions [13]. As a result of this, high energy shear waves and turbulence occur in the areas where cavitation occurs due to the increased temperature and pressure values [14].

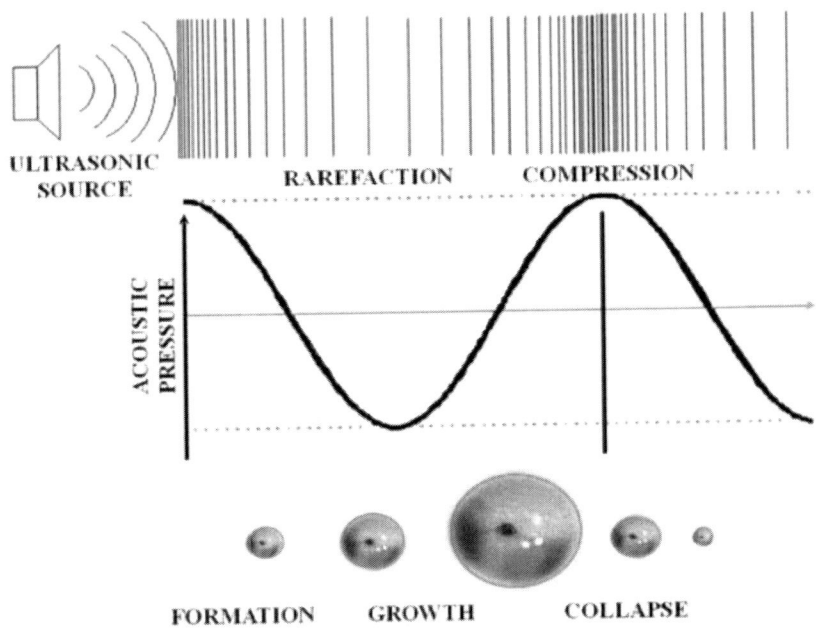

Figure 1. Cavitation.

Many factors affect the occurrence of cavitation phenomenon. At high temperatures, the vapor pressure rises and the surface tensile strength decreases, as a result of which more bubbles are formed. Although more bubbles appear at high temperatures, the intensity of the bubbles formed at the time of extinction is quite low. At low frequency values, the bubbles are produced in larger sizes and much more energy is generated if they are damped. If a high-intensity cavitation formation is requested, the size of ultrasonic sound waves should be increased [15].

The cavitation phenomenon finds application in 2 different ways, both temporary and stable:
- The process, which is described as temporary cavitation, takes place in probe systems at frequencies of 20 kHz and above and a wave intensity of 10 W/cm^2. In this unstable temporary cavitation process, the bubbles increase in volume with a small amount of acoustic oscillation and burst instantly. The resulting bubble sizes are small [11].
- In the case of stable cavitation, which occurs independently of the temporary cavitation event, the expansion of the bubbles occurs as a result of almost 1000 acoustic cycles at the frequency values of 200 kHz and above. The intensity of the explosions occurring in the bubbles is much lower when compared to the temporary cavitation [16].

Ultrasound System

The main elements of the ultrasound system are;

- generator,
- converters, and
- transmitting parts [16, 17].

While the generator device converts the alternating current to higher frequency alternating current; converters convert high frequency electric current into mechanical vibrations. These mechanical vibrations are transmitted to the environment where ultrasonic sound waves will take place in the transmitter section. There are two types of transducers in ultrasound systems, magnetostrictive and piezoelectric (Figure 2). The magneto strictive transducer acts as an electro-acoustic transducer in order to generate ultrasonic sound waves. On the other hand, piezoelectric transducers are concerned with the mutual conversion of acoustic and electrical energies. The main principle of a piezoelectric sensor or transducer is that when quartz crystals or any piezoelectric material is exposed to a force, it creates an electrical charge called piezoelectricity, which is responsible for the cleaning action on the surface.

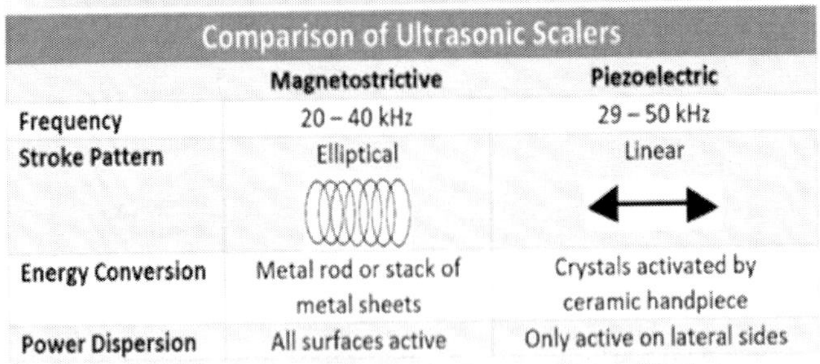

Comparison of Ultrasonic Scalers		
	Magnetostrictive	**Piezoelectric**
Frequency	20 – 40 kHz	29 – 50 kHz
Stroke Pattern	Elliptical	Linear
Energy Conversion	Metal rod or stack of metal sheets	Crystals activated by ceramic handpiece
Power Dispersion	All surfaces active	Only active on lateral sides

Figure 2. Piezoelectric and Magnetostrictive transducer.

Factors that play a decisive role in the application efficiency and effectiveness of ultrasonic sound waves:

- frequency,
- wavelength,
- amplitude and
- ultrasonic power.

In addition, it is possible to say that the reactor design and the realization of ultrasonic sound waves also affect the efficiency of the process.

The application of ultrasonic sound waves takes place in two different ways: ultrasonic bath and ultrasonic probe (Figure 3):

- Ultrasonic bath systems come to the fore in the removal of air present in solvents or the cleaning of small glass materials in general. Ultrasonic bath systems are rarely preferred in food processing streams due to the highly limited reproducibility of chemical reactions and low amounts of transferred energy density [18].
- As another method, it is possible to count the process in which ultrasonic sound waves are applied through ultrasonic probe (sonotrode) systems. In this method, since the probe is immersed in the sample, the ultrasonic intensity is transferred to a much more restricted region (only the tip of the probe). When we compare this system with the ultrasonic bath system, it will not be wrong to say that it is a much more effective method and finds use in small-scale

productions [19]. In case of cost comparison, it is possible to say that the bath system exhibits a much more affordable price range compared to the probe system.

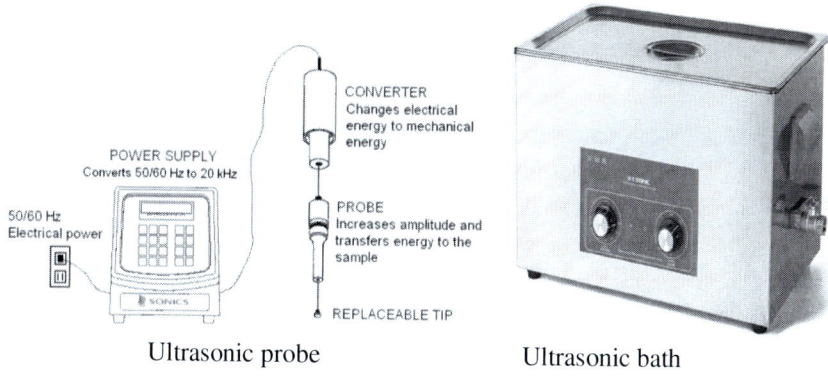

Ultrasonic probe   Ultrasonic bath

Figure 3. Ultrasound application method.

Usage Areas of Ultrasonic Sound Waves in the Food Industry

While ultrasonic sound waves reveal a sufficient level of efficiency alone in some processes, in some processes they require to be used together with factors such as temperature and pressure to reach a sufficient level of inactivation. These processes:

- ultrasonic sound waves + heat (thermosonication, TS),
- ultrasonic sound waves + pressure (manosonication, MS),
- ultrasonic sound waves + pressure + heat (manothermosonication, MTS) [13, 20].

It is used for pasteurization and sterilization purposes by reducing the process temperature and duration as a result of the application of thermosonication process, heat treatment and ultrasonic sound waves elements together. When the ambient temperature reaches levels close to the boiling point of the analyzed liquid, there is a decrease in the severity of cavitation with the rise of the liquid vapor pressure value. In order to overcome this problem, the thermosonication process is applied under pressure (100 kPa-700 kPa) and this process is expressed as manothermosonication. The effect resulting from the application of manothermosonication shows a synergistic character. That is, the resulting total effect is greater than the total of the effects produced by each separate factor (ultrasound, heat and pressure) that produces this effect if they are applied alone. In short, ultrasonic sound waves become more effective with the help of temperature and pressure parameters while sensitizing the target molecules and tissues [13].

Acoustic Drying

The process of removing water from a substance by means of a high-intensity sound field is called 'Acoustic Drying' [10]. Ultrasonic sound wave energy has a very suitable process for drying heat sensitive food products because it has a clear effect at low temperatures [21]. In the event that high-intensity ultrasonic sound waves directly affect the food material to be dried, it creates a series of compression and expansion movements within the material, such as the compression and release of a sponge material. The effect resulting from this mechanism makes it easier to remove water from the food material through the microscopic channels that occur as a result of the surface tension that traps the water in the food product, creating a much greater force [22]. In addition, the cavitation phenomenon that occurs with ultrasonic sound wave energy is thought to have an effective role in removing water [23].

Ultrasonic sound wave energy can be applied with processes such as:

- drying with hot air,
- by irradiation,
- by freeze drying,
- with high pressure,
- by osmotic drying [24].

In the drying process of food materials, ultrasonic sound wave energy has often been applied in combination with other drying methods. These processes have found application both simultaneously and as a pre-treatment with ultrasonic sound waves [25-27]. It has been determined that the pre-treatments used before the drying process in fruit and vegetable products reduce the long drying times and are effective in preventing drying problems such as low product quality [28]. Exposing the food materials to a pre-treatment without applying the drying process has become even more important recently, as it greatly affects the drying process. The application of pretreatment to food products creates a more porous texture in agricultural food products, increasing mass transfer rates during drying [29]. When fruits and vegetables are cut or sliced, it is possible to control the enzymatic browning reaction, which occurs through peroxides and various oxidative enzymes, in dried fruits and vegetables by pre-treatment. It has been determined that the pretreatment applied to food products shortens the drying time by 40%, reduces energy consumption and contributes to the preservation of bioactive components [30].

Osmotic Drying and Ultrasonic Sound Waves

Ultrasonic baths are mostly preferred in ultrasonic sound waves and osmotic drying process and this application takes place simultaneously. In the application of osmotic drying, a semi-permeable membrane with cellular surface effect assumes the responsibility. Water moves from the processed food material to the concentrated solution and the dissolved components move in the opposite direction [31]. In this case, ultrasonic sound waves are applied to remove water during osmotic drying and to increase the sugar gain. The reason for this is the food material originates from the microscopic channels created by the use of ultrasonic sound waves inside the capillaries, and thus moisture is removed easily [32].

In a study carried out on osmotic drying of apple fruit slices cut into cubes, as a result of the osmotic drying process in an ultrasonic bath system with 50 kHz frequency and 150 W energy, ultrasonic sound waves have an effect on mass transfer, water loss and sugar gain increase, at 40°C. It has also been shown that it can be achieved with a continuous mixing process without the application of ultrasonic sound waves [33].

Drying with Hot Air and Ultrasonic Sound Waves

In the drying process of food products with hot air system, an ultrasonic bath or probe methods can be applied as a pre-treatment before drying, or it can be carried out simultaneously with a probe positioned inside the dryer [34]. The aim of reaching the required humidity in the method of drying with hot air alone, which is carried out in fruits and vegetables, causes the drying time to be prolonged. This problem, which occurs due to poor mass transfer, causes hardening of the peels of fruits and vegetables dried by hot air. With the ultrasonic sound wave that takes place before the hot air-drying application, the water dispersion is increased to a great extent, shortening the time during the drying process and reducing the possible cost [35]. Ultrasonic sound waves applied as a pre-treatment significantly increase the mass transfer, water loss and solute amount in the drying process. The increase in mass transfer is effective in improving the drying rate of fruits and vegetables. In a study investigating the effects of ultrasonic sound waves on mushroom, carrot and apple products during hot air drying, it was stated that ultrasonic sound waves can greatly reduce the drying time by increasing dehydration rate [36].

Fernandes and Rodrigues [32] investigated the effects of ultrasonic sound wave pretreatment in the application of hot air drying of banana fruit as a result of a study they carried out. It is stated that the banana fruit, which was pretreated with an ultrasonic bath system with a frequency of 25 kHz at 10, 20 and 30 minutes and at 30°C, increased the water diffusion and reduced the total drying process by 10.3% as a result of the 20-minute process. In addition, as a result of the application of ultrasonic sound waves, the sugar content of banana fruits decreased by 21.3%. It has been revealed that ultrasonic sound wave application can enable to produce dried food products with a reduced sugar content [37].

In another study, it was reported that as a result of ultrasonic assisted hot air drying of cherry tomatoes, the diffusivity of water increased by 33-39% and the amount of carotenoids in dried food products could be maintained by applying ultrasonic sound waves [38].

In another study, it was found that there was an increase in the diffusion coefficient and mass transfer coefficients of the water during the application of ultrasonic sound wave assisted hot air drying of the strawberry fruit, this increase also decreased the increasing drying temperatures, and the ultrasonic sound wave treatment also decreased the total drying time at various drying temperatures. It has been reported to shorten it by 13% to 44% [39].

In another study investigating the effect of ultrasound energy on drying kinetics during hot air drying of potatoes, it was stated that the drying kinetics of potato slices cut into cubes showed a statistically significant difference with ultrasonic sound wave energy. It was determined that ultrasonic sound wave energy with a frequency of 21.8 kHz and a power of 37 kW/m^2 decreased the drying time by 40%, increased the diffusion coefficient of water by 64% and the mass transfer coefficient by 58% compared to the dried potato samples without ultrasound [40].

It was stated that the total flavonoid loss was 35.5% in apple samples using the hot air-drying method developed with ultrasonic sound waves, and it was at a much lower level than the flavonoid loss in apple fruits dried with the hot air method applied alone without ultrasonic sound waves [41]. It has been proven as a result of the studies that the drying speed can be shortened and the quality of the food product can be improved with the use of ultrasonic sound waves. In general, it is possible to say that ultrasonic sound waves are an effective method for maintaining both the total phenolic and total flavonoid ratios of the dried food products.

Spray Drying and Ultrasonic Sound Waves

During the drying of the starch component isolated from the raw banana fruit, in which ultrasonic sound waves were applied as a pre-treatment, the changes in both physical and chemical characteristics of the starch were examined. Ultrasonic sound waves were applied in the form of pretreatment at intervals of 1 min with the help of a vibrating probe with a frequency of 20 kHz and an energy of 24 W for an hour. Following this application, the drying was carried out with a mini spray dryer. As a result of the research, it was stated that the texture of starch molecules was disrupted by the effect of cavitation caused by the use of ultrasonic sound waves, water diffusion, swelling property of starch and water absorption capacity increased. In addition, in the same study, it was reported that the vibrations created by ultrasonic sound waves reduced the threshold shear stress and consistency coefficient of starch gels [39].

Freeze Drying and Ultrasonic Sound Waves

The problem experienced in freeze-drying application is the deterioration of the texture as a result of the formation of non-homogeneous crystal particles and a decrease in the sensory quality of the food product. In the freeze-drying application, in which ultrasonic sound waves are used as a pretreatment, thin uniform ice crystals are formed [19]. Cavitation bubbles created by ultrasonic sound waves act as nuclei for the growth of crystals. Ultrasonic sound waves create many nucleation sites in the water inside fruits and vegetables. The nucleation caused by ultrasonic sound waves consists of 2 steps as primary and secondary nucleation. These nuclei are separated into much smaller particles by the intense force that occurs as a result of the collapse of the cavitation bubbles [42]. Ultrasonic sound waves used as pre-treatment in food products cause a positive effect on the heat transfer coefficient during freezing [43].

Jambrak et al. [44] conducted a study in which ultrasonic sound waves were used for 3 to 10 min with an ultrasonic probe with a frequency of 20 kHz and an ultrasonic bath with a frequency of 40 kHz on vegetable samples of mushrooms, brussels sprouts and cauliflower. It was found that ultrasonic sound waves at 40 kHz frequency in the vegetables dried by hot air (60°C) and freeze-drying (-45°C) increase the drying rate in all of the samples except for the pre-treatment performed with an ultrasonic bath system for 10 min in freeze-dried. It has been determined that the rehydration characteristics of freeze-dried food products at low frequency are much higher [44].

Garcia et al. [45] investigated the effect of ultrasonic sound wave pre-treatment performed before freeze-drying on the color values of strawberry fruit. Considering the results of the research, it was determined that the $L*$ (brightness) and $b*$ (yellowness) parameters of the strawberry fruits that were subjected to ultrasonic sound wave pre-treatment were much fresher and had more reddish and bright colors than the strawberry samples that were not pre-treated with ultrasonic sound waves [45].

In the freezing mechanism supported by ultrasonic sound waves, 3 different categories of equipment are defined, which are full immersion, half immersion and non-immersion (Figure 4). In general, this equipment are basically consist of:

- ultrasonic system,
- refrigeration circulation combined with the refrigeration cycle, and
- temperature sensing sections combined with a data recording system.

Ultrasonic sound wave systems consist of a row of transducers connected to the generator with a power and frequency output evenly positioned at the bottom of a stainless ultrasound bath. The refrigerant circulation combined with the cooling system is pumped through the serpentine pipe located in the temperature control area, and the cooler at higher temperatures is cooled as it moves through the heat exchanger connected to the compressor.

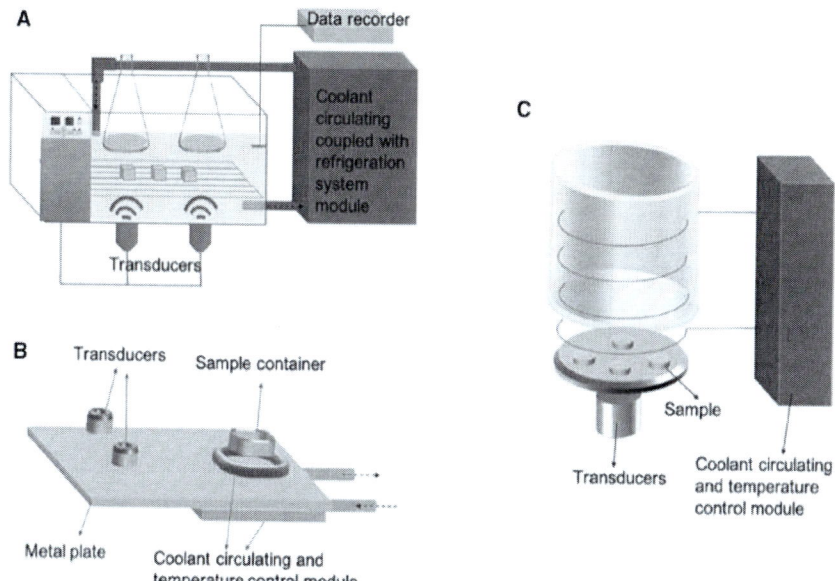

Figure 4. Types of ultrasound assisted freezing devices used in food processing: (A) Full immersion, (B) Half immersion, and (C) Non-immersion.

While the full immersion type equipment is frequently used in liquid, solid and semi-solid samples, the use of the other 2 types of devices, namely semi-immersion and non-immersion devices, is less frequently used. Semi-immersion type equipment is characterized by a metal plate tightly connected to ultrasonic transducers, and the samples (mainly liquid and semi-solid) in the vessel are placed on the tops of the plates in order to enhance the propagation of ultrasonic sound waves between the plate and the vessel. Finally, non-immersion devices allow direct contact between the ultrasound assisted freezing equipment and without immersion while the samples are cold. While this event prevents the contamination of the coolant, it also causes the loss of ultrasonic sound wave energy that occurs as a result of its propagation through the air.

Infrared Drying and Ultrasonic Sound Waves

Infrared radiation treatment, which has been focused on sensitively in the recent past, attracts a lot of attention in drying applications due to its significant thermal effects. This process increases the dehydration rate by penetrating both the surface and interior parts of the food in the drying application. Infrared drying application provides advantages in obtaining a quality food product with its drying speed when compared with traditional drying methods. In addition to food products, it is also used in medicine, plastic and paper industry, and it is very effective.

In a study examining the amount of bioactive substance of potato samples by contact ultrasound assisted infrared drying, it was determined that the amount of total phenolic substance increased and the phenolic component ratio increased proportionally as a result of increasing the ultrasonic power used. The easy oxidation and degradation of phenolic compounds causes a problem for drying applications. The use of ultrasonic sound waves with infrared application shortens the drying time and reduces the degradation of phenolic components.

It was determined that the drying time of sliced apple samples was significantly reduced in infrared drying with an ultrasonic sound wave pre-treatment [46]. In a study investigating the effects of ultrasonic sound wave energy on drying and textural characteristics of pear fruits in infrared drying application, pear fruit was exposed to ultrasound with an amplitude of 25%, 50%, 75% and 100% for 5 minutes with an ultrasonic probe with a frequency of 24 kHz, and then infrared drying process was carried out at 70°C. As a result, it has been revealed that ultrasonic sound wave pre-treatment shortens the drying time at increasing amplitudes due to the differences resulting from the cavitation phenomenon. In this study, as a result of the analyzes, it was determined that the hardness and elasticity decreased with increasing amplitude, and it was determined that the ultrasonic sound wave pre-treatment resulted with the best result in terms of texture [5].

Conclusion

Drying is among the oldest methods applied for the preservation of food products for a long time, and studies are still ongoing in terms of both final product quality and energy consumption with the help of advancing technology in the use of this process. While the drying application was done by drying the food product, which was traditionally placed in an open area, with natural convection, this method has been replaced by drying processes that provide faster, hygienic and homogeneous drying as time goes on, as a result of technological advances.

Ultrasonic sound wave technology, which has emerged as an alternative to thermal applications in the food industry, is applied in order to prevent microbial activity in food products and to increase the quality of the food product, as well as to preserve its physical, nutritional and organoleptic characteristics.

As a result of many studies, it has been determined that ultrasound technology is quite advantageous in the preservation of dried food products. Both ultrasonic sound waves alone and the combination of ultrasonic sound waves with different applications have positive effects in the application of drying food. Ultrasonic sound waves find successful use in many applications such as sterilization, filtration, emulsification, defoaming, apart from drying food products. With the widespread use of sustainable ultrasonic sound waves, positive effects can be created on the protection of the environment, mainly on foodstuffs.

References

[1] Abramov, O. V. (1998). *High-Intensity Ultrasound: Theory and Industrial Applications* London: Gordon and Breach.

[2] Earnshaw, R. G., Appleyard, J., ve Hurst, R. M. (1995), Understanding Physical Inactivation Proces: Combined Preservation Opportunites Using Heat, Ultrasound and Pressure. *International Journal of Food Microbiology*, 28(2), 197-219.

[3] Jambrak, A. R. (2012). Application of High-Power Ultrasound and Microwave in Food Processing: Extraction. *Journal of Food Processing and Technology*, 3-12.

[4] Dolatowski, Z. J., Stadnik, J., ve Stasiak, D. (2007). Applications of Ultrasound in Food Technology. *ACTA Scientiarum Polonorum – Technologia Alimentaria*, 6(3), 89-99.

[5] Jambrak, A. R., Herceg, Z., Subaric, D., Babic, J., Brncic, M., Brncic, S. R., Bosiljkov, T., Cvek, D., Tripalo, B., ve Gelo, J. (2010), Ultrasound Effect on Physical Properties of Corn Starch. *Carbohydrate Polymers*, 79, 91-100.

[6] Thakur, B. R., ve Nelson, P. E. (1997). Inactivation of Lipoxygenase in Whole Soy Flour Suspension by Ultrasonic Cavitation. *Molecular Nutrition*, 41(5), 299-301.

[7] Murphy, R., Beard, B., Macry, J., ve Berrang, M. E. (2009). *Application of Ultrasonic Technology for Killing Salmonella and Listeria monocytogenes in Fluid System*, Institute of Food Technology. June 6-9, 2009. Anaheim, CA. 236-01.

[8] Knorr, D., Zenker, M., Heinz, V., ve Lee, D. (2004). Applications and Potential of Ultrasonics in Food Processing. *Trends in Food Science and Technology*, 15(5), 261-266.

[9] Demirdöven, A., ve Baysal, T. (2009), The Use of Ultrasound and Combined Technologies in Food Preservation. *Food Reviews International*, 25, 1-11.

[10] Kantaş, Y. (2007). *Effect of Ultrasound on Drying Rate of Selected Produce*, Middle East Technical University, Natural and Applied Sciences, Ankara.

[11] Uzunoğlu, T. P. (2012). *Yüksek güçlü ultrases işleminin kısa ve uzun ömürlü ayranın mikrobiyolojik ve duyusal özelliklerine etkisi* (Yüksek Lisans Tezi). İTÜ Fen Bilimleri Enstitüsü, Gıda Mühendisliği Anabilim Dalı, İstanbul.

[12] Yildiz, G., ve Izli G. (2018). *Non-thermal process: Ultrasound Technology.* Saarbrücken, Germany: LAP LAMBERT Academic Publishing. ISBN: 978-3-659-97350-5.

[13] Rahman, S. M. (2007). *Handbook of Food Preservation.* (2nt Ed.). CRC Press. London. 713-739.

[14] Patist, A., ve Bates, D. (2008). Ultrasonic innovations in the food industry: From the laboratory to commercial production. *Innovative Food Science and Emerging Technology*, 9, 147-154.

[15] Tüfekçi, S., ve Özkal S. G. (2015). Gıdaların Kurutulmasında Ultrases Kullanımı. *Pamukkale Üniversitesi Mühendislik Bilimleri Dergisi*, 21 (9), 408-413.

[16] Türkmen, F. (2012). Yüksek güçlü ultrases işleminin sütün fizikokimyasal ve jelleşme özelliklerine etkisi (Yüksek Lisans Tezi) *İTÜ Fen Bilimleri Enstitüsü*, Gıda Mühendisliği Anabilim Dalı, İstanbul.

[17] Mason, T. J. (1998). Power Ultrasound in Food Processing-The Way Forward: In *Ultrasound in Food Processing.* Blackie Academic & Professional, London, s. 105-126.

[18] Wen C., Zhang J., Zhang H., Dzah CS., Zandile M., Duan Y., ve Luo X. (2018). Advances in ultrasound assisted extraction of bioactive compounds from cash crops- A review. *Ultrasonic Sonochemistry*, 48, 538-549.

[19] Chemat, F., Zill-e-Huma, ve Khan, M. K. (2011). Applications of Ultrasound in Food Technology: Processing, Preservation and Extraction. *Ultrasonics Sonochemistry*, 18, 813-835.

[20] McClements, D. J. (1995). Advances in the Application Ultrasound in Food Analysis and Processing. *Trends in Food Science and Technology*, 6, 293-299.

[21] Rastogi, N. K. (2011). Opportunities and challenges in application of ultrasound in food processing. *Critical Reviews in Food Science and Nutrition*, 51, 705-722.

[22] De La Fuente-Blanco, S., Riera-Franco de Sarabia, E., Acosta-Aparicio, V. M., ve Gallego-Juarez, J. A. (2006). *Food Drying Proces by Power Ultrasound*, Ultrasonics, 44, 523-527.

[23] Garcia J., Oliveira, F. I. B., Weller, C. L., Rodrigues, S., ve Fernandes, F. A. N. (2014). Effect of Ultrasonic and Osmotic Dehydration Pre-treatments on the Colour of Freeze-Dried Strawberries. *Journal of Food Science and Technology*, 51(9), 2222-2227.

[24] Topdaş, E. F., Başlar, M., ve Ertugay, M. F. (2011). Elmaların ozmotik kurutulması üzerine ultrases işleminin etkisi. *Akademik Gıda*, 9(5), 6-10.

[25] Yildiz, G., Rababah, T., ve Feng, H. (2016). Ultrasound- Assisted Cutting of Cheddar, Mozzarella and Swiss Cheeses – Effects on Quality Attributes during Storage. *Innovative Food Science and Emerging Technologies*, 37, 1-9.

[26] Yildiz, G., ve Izli, G. (2019). The effect of ultrasound pretreatment on quality attributes of freeze-dried quince slices: Physical properties and bioactive compounds. *Journal of Food Process Engineering*, 42 (5), e13223.

[27] İzli, G., ve Yildiz, G. (2021). Evaluation of the high intensity ultrasound pretreatment effects on the physical properties and bioactive compounds of convective dried quince samples. *International Journal of Fruit Science*, 21(1), 645-656.

[28] Yildiz, G. (2021). The Effect of High Intensity Ultrasound Pre-treatment on the Functional Properties of Microwave-dried Pears (Pyrus communis). *Latin American Applied Research Journal*, 51(2), 133-137.

[29] Ahrné, L., Prothon, F., ve Funebo, T. (2003). Elma ve patatesin mikrodalga destekli dehidrasyonundan önce iki kalsiyum ön işleminin kurutma kinetiği ve doku etkilerinin karşılaştırılması. *Uluslararası Gıda Bilimi ve Teknolojisi Dergisi*, 38, 411–420.

[30] Yıldız, G., İzli, G., Çavuş, M., ve Ceylan, M. M. (2021). Ultrason Ön İşleminin Kurutulmuş Iğdır Kayısısının Kalite Özellikleri Üzerine Etkisi. *Journal of the Institute of Science and Technology*, 11(1), 303-313.

[31] Nahimana, H., Zhang, M., Mujumdar, A. S., ve Ding, Z. (2011). Meyve ve sebzelerin ozmotik dehidrasyonu sırasında kütle transfer modellemesi ve büzülme düşüncesi. *Food Reviews International*, 27 (4), 331-356.

[32] Fernandes, F. A. N., ve Rodrigues, S. (2007). Meyvelerin kurutulması için ön işlem olarak ultrason: Muzun dehidrasyonu. *Gıda Mühendisliği Dergisi*, 82, 261-267.

[33] Simal, S., Benedito, J., ve Rosello, C. (1998). Use of ultrasound to Increase Mass Transport Rates during Osmotic Dehydration. *Journal of Food Engineering*, 36(3), 323- 336.

[34] Yıldız, G., ve İzli G. (2020). *Microwave and Convective Drying Methods on Food Drying*. Ankara, Turkey: Iksad International Publishing House. ISBN: 978-605-7811-87-5.

[35] Oliviera, F. I. P., Gallão, M. I., Rodrigues, S., ve Fernandes, F. A. N. (2010). Dehydration of malay apple (Syzygium malaccense L.) using ultrasound as pretreatment. *Food Bioprocess Technology*, 4, 610-615.

[36] Gallego-Juárez, J. A., Riera, E., de la Fuente Blanco, S., Rodríguez-Corral, G., Acosta-Aparicio, V. M., ve Blanco, A. (2007). Application of High-Power Ultrasound for Dehydration of Vegetables: Processes and Devices, *Drying Technology*, 25(11), 1893-1901.

[37] Fernandes, F. A. N., Gallão, M. I., ve Rodrigues, S. (2009). Effect of osmosis and ultrasound on pineapple cell tissue structure during dehydration. *Journal of Food Engineering*, 90, 186-190.

[38] Fernandes F. A. N., Rodrigues, S., García-Pérez, J. V., ve Cárcel, J. A. (2015). Ultrason Destekli Hava Kurutmanın Kiraz Domateslerin Vitaminleri ve Karotenoidleri Üzerindeki Etkileri. *Drying Technology*, 34, 986–996.

[39] Izidoro D. R., Sierakowski M. R., Haminiuk CWI., Souza C. F., ve Scheer A. G. (2011). Physical and Chemical Properties of Ultrasonically Spray-Dried Green Banana Starch. *Journal of Food Engineering*, 104(4), 639-648.

[40] Ozuna, C., Carcél, J. A., Garcia-Perez, J. V., ve Mulet, A. (2011). Improvement of water transport mechanisms during potato drying by applying ultrasound. *Journal of Science Food and Agriculture*, 91, 2511-2517.

[41] Rodríguez, J., Mulet, ve A., Bon, J. (2014). Influence of high-intensity ultrasound on drying kinetics in fixed beds of high porosity. *Journal of Food Engineering*, 127, 93-102.

[42] Xu, H. S., Zhang, M., Duan, X., Mujumdar, A. S., ve Sun, J. C. (2009). Dondurarak kurutmadan önce edamame üzerinde güçlü ultrason ön işleminin etkisi. *Kurutma Teknolojisi*, 27(2), 186-193.

[43] Zheng, L., ve Sun, D-W., (2006). Innovative Applications of Power Ultrasound During Food Freezing Processing-a Review. *Trends in Food Science & Technology*, 17, 16-23.

[44] Jambrak, A. R., Mason, T. J., Paniwyk, L., ve Lelas, V. (2007). Accelerated Drying of Button Mushrooms, Brussels Sprouts and Cauliflower by Applying Power Ultrasound and its Rehydration Properties. *Journal of Food Engineering*, 81(1), 88-97.

[45] Garcia-Perez, J. V., Carcél, J. A., Benedito, M., ve Mulet, A. (2007). Power ultrasound mass transfer enhancment in food drying. *Food and Bioproducts Processing Journal*, 85, 247-254.

[46] Brncic, M., Karlovic, S., Rimac Brncic, S., Penava, A., Bosiljkov, T., Ježek, D., ve Tripalo, B. (2010). Kızılötesi kurutulmuş elma dilimlerinin yüksek güçlü ultrason ön işleminden etkilenen dokusal özellikleri. *African Journal of Biotechnology*, 9(41), 6907-6915.

Chapter 3

The Effect of Sonication Treatment on the Quality Properties of Apple-Carrot Juice

Pınar Altaş and Gökçen Yıldız[*]

Department of Food Engineering, Bursa Technical University, Bursa, Turkey

Abstract

In current study, a mixture of apple and carrot juice prepared at a ratio of 60:40 was processed with traditional pasteurization (PAS) at 90°C for a min, sonication (US) for 30 min and thermosonication (TS) at 40°C, 50°C and 60°C for 30 min. The collected apple-carrot juice mixture was stored at both room temperature (25±0.5°C) and cold storage (4±0.5°C) for 28 days. The quality and microbiological analyses of control (untreated), PAS, US and TS-treated samples were performed on days 0, 14 and 28.

The thermosonicated samples resulted with a significant inactivation of aerobic mesophilic bacteria and yeast-molds were compared to the untreated samples. Number of bacteria and yeast-mold growth was not observed in TS-50°C and TS-60°C samples, and it was determined that the microbiological shelf life of these samples continued and they were consumable after storage. Quality characteristics such as acidity, pH, brix, browning index and color (L^*, a^*, b^*, ΔE) were not changed with the use of sonication treatments. Total phenolics, antioxidant capacity, total carotenoids, ascorbic acid remained at higher concentration in the sonicated samples compared to the pasteurized samples. Lower level of Hydroxymethyl furfural (HMF) were obtained with the sonication treatments. Applying sonication has a positive impact on color,

---

[*] Corresponding Author's Email: gokcen.yildiz@btu.edu.tr.

In: Power Ultrasound and Its Applications in Food Processing
Editors: Gulcin Yildiz and Gökçen Yıldız
ISBN: 979-8-88697-639-7
© 2023 Nova Science Publishers, Inc.

appearance, odor, flavor and general taste compared to thermal treatments. Thus, considering all of these aspects, TS at 50°C and 60°C might be used as an alternative to heat treatment, further it can be successfully applied in the beverage industry for the apple-carrot juice production.

Keywords: apple-carrot juice, pasteurization, shelf-life, sonication, thermosonication

Introduction

Fruits and vegetables are very important for a healthy diet, as they contain high amounts of dietary fiber, vitamins, minerals and phytochemicals such as antioxidants, phytoestrogens and anti-inflammatory [1]. In recent years, interest in fruit juices obtained by mixing different fruits has been increased. In addition to improving taste and aroma, mixed juices provide enrichment of bioactive compounds, nutrients, flavor and appearance, and reduce production costs as well [2]. The apple is one of the healthy and nutritious foods because it contains antioxidant substances such as monosaccharides, minerals, organic acids, dietary fiber, phenolic compounds, and vitamins A, B1, B2, C and E [3].

Carrot (*Daucus carota L.*) juice has important nutritional sources such as α- and β-carotene, zeacaroten, lutein and lycopene. Beta-carotene is one of the most biologically effective carotenoids and is known as a pro-vitamin A. Carrot is a good source of carbohydrates and minerals such as calcium, phosphorus, iron and magnesium [4]. Besides the nutritional value of carrot juice, its acidity is low and its pH value is around 6, which creates a favorable environment for the growth of many spoilage and spore-forming bacteria. Acidification of carrot juice naturally blending carrot juice with acidic juices such as apple juice (pH 3.5-4.0) can be produced with a mixture with a lower pH that can create a natural barrier against most microorganisms [5].

Conventional pasteurization (PAS), which is a heat treatment, is used to prevent the spoilage of fruit and vegetable juices result from microbial growth and enzymatic activity. However, while heat treatment preserves the shelf life and stability of fruit and vegetable juices, it can cause negative changes in quality properties such as phenolic substances, volatile components, vitamins (especially C and E), carotenoids, anthocyanins, organic acids and color in fruit juices [6].

Sonication (US), which have been proven for their potential to be used in the food industry, especially in the beverage industry, is used as one of the innovative technology alternatives to full or partial heat [6]. Sonication is mechanical energy consisting of sound waves that generate 20.000 or more vibrations or oscillations per second [7]. The inactivation effect of sonication occurs with the physical and chemical effects of acoustic cavitation, which is described as the formation, growth and implosion of bubbles by sound waves in a liquid medium [8]. Acoustic cavitation occurs when the ultrasonic wave passes through a liquid medium, promoting alternate regions of compression and rarefaction. Thus, it leads to the formation of bubbles in the liquid medium. If cavitation is unstable; there is an implosion of the bubbles and causes a temperature of up to 5500°C, a pressure of 50 MPa [8-10] a high rate of heating, cooling (10 9 K/s) and shock waves are formed at a very short time. The application of US in combination with temperature, pressure or both increases the efficiency of the US process [11]. Thermosonication (TS), which is the combination of US and heat, increases enzyme and bacterial inactivation without affecting the juice quality [8].

High-energy low-frequency (20 kHz -100 kHz) sound waves are called "power ultrasound" and these waves form a cavitation which inactivate microorganisms [7, 9]. Sonication can be applied either with heat or pressure, or both. The use of US with heat is called 'thermosonication', its use with pressure alone is called 'manosonication', and its use with both temperature and pressure is called 'manothermosonication' [11].

The bactericidal effect of US is caused by the intracellular cavitation. Intracellular micromechanical shocks disrupt cellular, structural and functional components and cause the cells to break down [8].

The presence of endogenous enzymes such as pectinmethylesterase (PME), polyphenoloxidase (PPO), peroxidases (POD) and lipoxygenase (LOX) in fruits can affect the quality of these products in different ways, such as loss of viscosity during processing of fruit juices, bad taste and formation of browning pigments. Enzyme inactivation occurs by protein denaturation due to the formation of free radicals during the sonolysis of water molecules, the formation of cavitating bubbles or shear forces from the collapse of the bubbles [12].

The presence of bioactive compounds such as ascorbic acid, carotenoids and phenolic compounds in fruit and vegetable juices determines the nutritional value of these products. Therefore, the effect of US treatments on these compounds is of a great importance. Cavitation is directly responsible for the physical conditions that occur as a result of cavitation precipitation on the quality properties of fruit juices. It is stated that heat and cavitation help remove dissolved oxygen from the environment and thus delay ascorbic acid degradation. In addition, it is reported that hydroxyl radicals formed during cavitation can easily react with oxidizable foods and cause ascorbic acid degradation [6]. As a result of occurrence of damage in the cell wall by US, the extraction of phenolic and carotenoid compounds increases. The increase of these compounds increases the antioxidant capacity in the medium [6, 13]. However, as a result of free radicals formed by cavitation, antioxidant compounds reduce their concentrations in the medium by acting on free radicals [14].

Some studies have shown that the TS treatment applied to tomato juice [15], pineapple, grape, cranberry juices [16], apple juice [17], mango juice [18], carrot juice [19], pear juice [20] and blueberry juice [21] decreased microbial counts without significantly affecting the physicochemical properties of the juices. Sonication treatments also showed positive effect on the release of bioactive compounds and increase antioxidant activity of melon juice [22], grapefruit juice [23], mango juice [18], strawberry juice [24, 25], carrot juice [19, 26], peach juice [27], and cherry juice [28].

In this study, it is aimed to ensure the microbial safety by pasteurization and eliminate the negative effect of pasteurization on fruit juice quality because of high temperature with US and TS treatments with moderate temperature (lower than 60°C) as an alternative to conventional PAS.

Material and Method

Material

In the study, granny apples and carrots were purchased from a local market in Bursa, Turkey. Apples and carrots were stored at 4°C until the juice production. Mixed fruit juice was obtained after the squeezing and filtering the apples and carrots with a juicer (Vestel/Energic KMS 6000, Turkey).

Chemicals

Phenolphthalein indicator solution, acetone, sodium carbonate, Folin-Ciocalteu solution, DPPH (2,2-Diphenyl-1-picrylhydrazyl) and HMF standard from Sigma-Aldrich (St. Louis, MO, USA); sodium hydroxide, hexane and methanol from ISOLAB (Wertheim, Germany); a solution of ascorbic acid, metaphosphoric acid and 2.6 dichlorophenolindophenol from Merck (Damstadt, Germany); Carrez I and Carrez II solutions from Carl ROTH (Karlshure, Germany) and ethyl alcohol from Tekkim (Bursa, Turkey) were used in the analysis of the research.

Treatments and Apple-Carrot Juice Production

A mixture of apple and carrot juice was prepared at a ratio of 60:40. Conventional PAS at 90°C for a minute, US at 53 kHz frequency, 160 W power, 100% amplitude for 30 min at room temperature, and TS at 40°C, 50°C, 60° for 30 min with a frequency of 53 kHz, power of 160 W and 100% amplitude were applied to the apple-carrot juice mixture. After the treatments, apple-carrot juices were stored at room temperature (25±0.5°C) and refrigerator (4±0.5°C) for 28 days. Control (untreated), PAS, US and TS were applied to the juice samples. The quality and microbiological analyzes of the untreated and treated samples were carried out on days 0, 14 and 28. The process conditions applied to apple-carrot juice is shown in Table 1.

Table 1. Process conditions applied to the apple-carrot juice

Samples	Treatments	Instrument characteristics
Control	Untreated fresh apple-carrot juice (60:40) mix	-
Pasteurization (PAS)	Apple-carrot juice (60:40) pasteurized at 90°C for a min	Termal-J11330 KD (Turkey)
Sonication (US)	Apple-carrot juice (60:40) sonicated for 30 min at room temperature	Kudos-SK3310 HP (China)
Termosonication-40°C (TS-40°C)	Apple-carrot juice (60:40) thermosonicated for 30 min at 40°C	Kudos-SK3310 HP (China)
Termosonication-50°C (TS-50°C)	Apple-carrot juice (60:40) thermosonicated for 30 min at 50°C	Kudos-SK3310 HP (China)
Termosonication-60°C (TS-60°C)	Apple-carrot juice (60:40) thermosonicated for 30 min at 60°C	Kudos-SK3310 HP (China)

Methods

Microbiological Analyzes

Total mesophilic aerobic bacteria were determined in PCA (Plate Count Agar) medium and total yeast mold was determined in PDA (Potato Dextrose Agar) medium.

Total Mesophilic Aerobic Bacteria Count
1 mL of apple-carrot juice mixture was taken from the samples and transferred to tubes containing 9 mL of FTS, and 1 mL was taken from the prepared dilution tubes and inoculated into PCA medium by pouring method. The results after 48 hours of incubation at 37±2°C are given as cfu/mL [29].

Total Yeast and Mold Count
1 mL of apple-carrot juice mixture was taken from the samples and transferred to tubes containing 9 mL of FTS, and 100 µL was taken from the prepared dilution tubes and inoculated into PDA medium by spreading method. The results after 5 days of incubation at 22±2°C are given as cfu/mL [29].

Determination of Soluble Solids (°Brix)
Brix values indicating soluble dry matter content were measured with a digital refractometer (Kyoto KEM/RA-600, Japan) at a standard degree (20°C) [30].

Determination of pH
The pH values of juice samples were measured with a glass electrode digital pH-meter (Ohaus Starter 3100 model, USA) at 20°C [30].

Determination of Titratable Acidity
5 mL of apple-carrot juice mixture was diluted with distilled water at a ratio of 1:10 and transferred to a flask. 0.25-0.5 mL of 1% phenolphthalein indicator solution (Sigma, Germany) was added and titrated with 0.1 N NaOH (Isolab, Germany) solution for 30 s until a permanent light pink color was obtained [31]. Results calculated as % acidity in citric acid (2.1).

$$\%\text{Titratable Acidity} = (V \times meq \times 100)/S \tag{2.1}$$

V= Titer volume of 0,1 N NaOH solution (mL)
N= Normality of NaOH
F= Factor of NaOH
S= Amount of juice sample (mL)
meq= Milliequivalent weight of organic acid (citric acid: 0.064)

Extraction for Total Phenolic Content and Antioxidant Capacity
1.5 mL apple-carrot juice sample was taken into 15 mL falcon tubes and 2.5 mL 80% methanol was added. After mixing the tube contents at 200 rpm for

15 minutes in an orbital shaker (Mipro/MLS3535, Turkey), the samples were passed through a 0.45 μm PTPE membrane filter.

Determination of Total Phenolic Content

The total amount of phenolic content was determined by the Folin-Coicalteau method. The principle of the Folin-Ciocalteu method is based on the reduction of phenolic compounds in the Folin-Ciocalteu solution in a basic medium. After 0.2 mL of extract was taken into a 15 mL falcon tube, 1.5 mL of 1:10 diluted Folin-Ciocalteu solution was added. The sample was incubated for 5 minutes after vortexing (Velp Scientifica F202 A0173, Italy). After adding 1.5 mL of 7% Na_2CO_3 solution, the mixture was vortexed again and kept in the dark for 90 minutes. The absorbance of the blue color formed by the reduced Folin-Ciocalteu solution as a result of the reaction was read in the spectrophotometer (Thermo Scientific EvolutionTM 201, USA) at a wavelength of 765 nm. Total phenolic content of apple-carrot juice samples was evaluated using the prepared gallic acid (GA) (5-50 mg/L) calibration curve and calculated in mg GAE/L (y= 0.00088 x + 0.003, $R2=0.97832$) [32].

Determination of Antioxidan Capacity

DPPH (2,2-Diphenyl-1-picrylhydrazyl) free radical scavenging activity method was used to determine the antioxidant capacity of apple-carrot juice samples. This method is associated with the reduction of the DPPH radical, which is a stable and purple colored compound, by antioxidant compounds. The purple-colored DPPH radical shows a strong absorption at a wavelength of 515-517 nm, which can be easily detected in the spectrophotometer. The method is basically based on measuring the decrease in the purple color of the radical solution, which occurs as a result of adding the test compound to the DPPH radical solution prepared in methanol or ethanol, in a spectrophotometer at a wavelength of 515-517 nm [31].

0.1 mL of the prepared extract was taken into a 15 mL falcon tube and 3.9 mL of 0.1 mM DPPH (Aldrich, Germany) solution prepared with ethanol was added. After the samples were vortexed (Velp Scientifica F202 A0173, Italy), they were kept in the dark for 30 min and the absorbance was read with a spectrophotometer (Thermo Scientific EvolutionTM 201, USA) at 515 nm wavelength. For the blank sample, the same system was achieved using a methanol instead of a sample. From the results, antioxidant capacity levels were calculated as μmol Troloks/mL with the help of the prepared trolox curve (0.1-1.0 mM) (y= 81.75 x – 12,627, $R2=0.9886$).

Determination of Total Carotenoids

Total carotenoid content of samples were measured by spectrophotometric method by using the modifying the method used by Lee et al. [33]. 3 mL apple-carrot juice sample was taken into 15 mL falcon tubes and 6 mL extraction solution (hexane: acetone: methanol, 50:25:25, v/v + 0.1 g ascorbic acid) was added and the tubes were vortexed. Samples were mixed in an orbital shaker (Mipro/MLS3535 model, Turkey) at 200 rpm for 30 min and then centrifuged at 9000 rpm at 4°C for 10 min (Hettich/Universal 320 R, Germany). The absorbance of the upper clear phase, quickly removed with a pasteur pipette, was measured in a hexane calibrated spectrophotometer (Thermo Scientific EvolutionTM 201, USA) at a wavelength of 450 nm.

The calculation was achieved using the β-carotene extinction coefficient $\varepsilon=2505$ according to De Ritter and Purcell [34]:

$$A \times 10000 = \varepsilon \times b \times c \tag{2.2}$$

A= Absorbance
ε= 2505 (β-karoten)
b= 1
c= mg/L

Results were given in mg/L as β-carotene.

Determination of Ascorbic Acid

For the determination of ascorbic acid in apple-carrot juice samples, 2.6 dichlorophenolindophenol titration method was applied. 5 mL of apple-carrot juice mixture was diluted with 3% metaphosphoric acid solution at a certain rate, mixed homogeneously and filtered through a filter paper. The obtained filtrate was titrated with a solution of 2,6-dichlorophenolindophenol until a pink color was formed and this color remained constant for 15 seconds. From the results, the amount of ascorbic acid (mg/100 mL) was calculated with the below equation (2.3) [30]:

$$\text{Ascorbic acid (mg/100 mL)} = [V \times F \times 100]/M \tag{2.3}$$

V: Volume of 2,6-diklorofenolindofenol solution used for titration (mL)
F: Amount of ascorbic acid equivalent to 1 mL of 2,6-dichlorophenolindophenol solution (mg)
M: Original sample amount in filtrate taken for titration (mL)

Determination of Hydroxymethyl furfural (HMF)

Hydroxymethyl furfural (HMF) determination was achieved by chromatographic method. The characteristics and conditions for the HPLC instrument is given in Table 2.

Table 2. The characteristics and conditions of HPLC

Instrument	HPLC (Agilent/İnfinity 1260)
Software	Lab Advisor Software
Detector	Diode array detector (DAD)
Column	C18 (150 mm x 4.6 mm, 5 μm)
Column temperature	30°C
Mobile phase A	%1 acetic acid aqueous solution (%90)
Mobile phase B	Methanol (%10)
Flow rate	0.8 mL/min
Injection volume	10 μL
Wavelength	280 nm
Retention time	5.8 min

A sample of 10 mL apple-carrot juice was taken and vortexed by adding Carrez I and Carrez II solutions (Velp Scientifica F202 A0173, Italy). After waiting for 10 min, the extracts, which were centrifuged at 9000 rpm for 10 min at 4°C (Hettich/Universal 320 R, Germany) and passed through a 0.45 μm PTPE membrane filter, were taken into vials and measured in HPLC (Agilent Infinity 1260). Results were calculated using the calibration curve prepared with standard solutions (0.1-10 mg/L) (y=102.44 x − 11,584, R^2=0.9961) and shown as mg/L.

Determination of Browning Index

5 mL of ethyl alcohol was added to 5 mL of apple-carrot juice, and the content of the tube was centrifuged at 9000 rpm for 10 min at room temperature (24°C) with a centrifuge (Hettich/Universal 320 R model, Germany). The absorbance of the upper clear part, which was quickly taken after centrifugation, was measured with a UV/VIS spectrophotometer (Thermo Scientific EvolutionTM 201, USA) calibrated with a pure water at a wavelength of 420 nm [35].

Color Measurement

Color analysis of the samples was performed using a Konica Minolta (CR-400, Japan) color measuring device according to the CIE (International Commission on Illumination) color measurement system. The L^* axis shows the change in brightness (L=0 black; L=100 white), the a^* axis shows the scale from green to red (-a=green, +a=red), the $b*$ axis shows the change from yellow to blue (-b). =blue; +b=yellow). Since the $L*$, $a*$ and $b*$ values are not perceived directly by the customers, the $ΔE$ value, which gives information about the color change, was also calculated. The equations to be used in the calculation of $ΔE$ values are given below [32]. L_0*, a_0* ve b_0* values used in the equation are the color values of first day of the apple-carrot juice obtained after the treatments.

$$ΔE = \sqrt{(L*-L_0*)^2 + (a*-a_0*)^2 + (b*-b_0*)^2} \qquad (2.4)$$

Sensory Analysis

Sensory analysis of apple-carrot juice samples was carried out by a trained panelist group of 17 people. The samples were coded with three digital numbers and water/grissini was presented to the panelists between the samples to avoid interaction between the samples. Panelists were asked to make a sensory evaluation using a 1-5 hedonic scale (1-Bad, 2-Not enough, 3-Acceptable, 4-Good, and 5-Very good) by giving information to the panelists on how to evaluate the samples in terms of color, appearance, smell, taste and general taste.

Statistical Analysis

Experiments were carried out in 2 replications. In the statistical evaluation of the obtained data, variance analysis was performed using JMP Statistical Discovery Software 7.0 package program (SAS Institute Inc., Cary, USA) and the least significant difference (LSD) multiple comparison test was applied at $α=0.95$ confidence interval.

Results and Discussion

Changes in Total Aerobic Bacteria and Yeast-Mold Counts during Storage

The changes of apple-carrot juice during 28 days of storage on microbial load are given in Table 3. The total number of mesophilic aerobic bacteria of the control sample was found as 1.0×10^4 cfu/mL and the total number of yeast molds as 4.5×10^3 cfu/mL. In the US sample, the initial total mesophilic aerobic bacteria count was 7.5×10^2 cfu/mL, and the total yeast mold count was 7.0×10^2 cfu/mL. After PAS and TS (TS-40°C, TS-50°C, TS-60°C) treatments, no bacteria and yeast-mold growth was observed in the samples (Table 3).

The treated samples provided significant inactivation in the total number of aerobic mesophilic bacteria and yeast-molds compared to the untreated sample on day 0. During 28 days of storage, microbial growth was not observed in the samples of PAS and TS-40°C, TS-50°C, TS-60°C treatments as well. However, in the TS-40°C sample stored for 28 days at 25°C, the total number of mesophilic bacteria and yeast-mold were found to be 3.0×10^3 cfu/mL and 3.8×10^2 cfu/mL, respectively. The microbial growth was the highest in control sample after 28 days at 25°C of storage, while other treatments showed less growth of microorganisms. The US treatment was effective in reducing the initial microbial count compared to the untreated sample. The microbial load was increased with the increase in storage time and temperature. This findings similar to those observed in apple-carrot juices [5] and carrot juice [36, 37] during storage.

Table 3. Microbiological analysis of apple-carrot juice samples processed with various treatments during storage

Storage time	Treatments	Total mesophilic aerobic bacteria count (cfu/mL)		Total yeast-mold count (cfu/mL)	
		Storage temperature		Storage temperature	
		4°C	25°C	4°C	25°C
Day 0	Fresh	1.0×10^4	1.0×10^4	4.5×10^3	4.5×10^3
	PAS	-	-	-	-
	US	7.5×10^2	7.5×10^2	7.0×10^2	7.0×10^2
	TS-40°C	-	-	-	-
	TS-50°C	-	-	-	-
	TS-60°C	-	-	-	-
Day 14	Fresh	5.2×10^5	3.6×10^6	2.0×10^4	1.8×10^5
	PAS	-	-	-	-
	US	3.8×10^3	5.3×10^4	2.8×10^3	4.5×10^4
	TS-40°C	-	-	-	-
	TS-50°C	-	-	-	-
	TS-60°C	-	-	-	-
Day 28	Fresh	1.0×10^9	1.5×10^{10}	6.0×10^5	1.0×10^7
	PAS	-	-	-	-
	US	5.5×10^5	1.0×10^6	2.0×10^4	2.6×10^5
	TS-40°C	-	3.0×10^3	-	3.8×10^2
	TS-50°C	-	-	-	-
	TS-60°C	-	-	-	-

In current study, the results of 28 days of storage period of apple-carrot juice samples at room temperature (25±0.5°C) and refrigerator (4±0.5°C) conditions, except US (14th and 28th day samples) and TS-40°C (28th days at 25°C) samples, it is seen that the shelf life of the processed samples continues and they are at a consumable level after storage. For the control sample, the microbial load after 14 days of storage is not at an acceptable level. For this reason, while physicochemical analyzes performed on apple-carrot juice samples were conducted only on day 0 in control and US samples, it was performed on TS-40°C application except for the 28th day sample stored at 25°C.

The inactivation effect by heat treatment is explained by disrupting the cell membrane and causing cytolytic effects by damaging nucleic acids. Microbial inactivation by TS occurs with the mechanical effect of cavitation. Bubbles collapsing by cavitation cause local temperature and pressure increase, leading to deterioration of the cell wall structure of microorganisms. In addition, the formation of free radicals (OH and H+) and hydrogen peroxide as a result of the sonolysis of water inactivates microbial cells [12]. The microorganisms are most sensitive to sonication occurs at temperatures above 55°C [38]. Sonication treatments effect on microbial inactivation but sonication with heat (thermosonication) provide a greater effect for the inactivation of microorganism [33].

Brix, pH, Titratable Acidity

The effects of treatments on °brix values, pH and acidity of apple-carrot juice samples determined during 28 days of storage are displayed in Table 4. According to Table 4, average brix values of all treated samples were found in the range of 9.97±0.91–11.10±0.10. There was no significant ($p>0.05$) difference was observed between the average brix values of the all treatments applied to the apple-carrot juice mixture during 28 days of storage compared to the control. Similar results were also observed by Zia et al. [39] who studied sonication-microwawe treated sugarcane juice during cold storage. However, brix values were affected by storage time and temperature caused a decrease in brix values ($p<0.05$). The lowest brix content of the samples might be due to spoilage and fermentation of the juices as a result of the conversion of sugars to acids, carbon dioxide, or alcohol [40]. There was no significant changes ($p>0.05$) in pH values in the all treatments during 28 days of storage compared to the control sample.

Table 4. Brix, pH and titratable acidity properties of apple-carrot juice samples processed with various applications during storage

		Brix	pH	Titratable acidity
Treatments	Fresh	10.70 ± 0.10[abc]	3.77 ± 0.00[ab]	0.52 ± 0.00[a]
	PAS	10.11 ± 0.60[bc]	3.81 ± 0.08[ab]	0.48 ± 0.04[abc]
	US	11.10 ± 0.10[a]	3.74 ± 0.00[b]	0.52 ± 0.01[ab]
	TS-40 °C	10.74 ± 0.55[ab]	3.93 ± 0.12[a]	0.45 ± 0.02[c]
	TS-50 °C	9.97 ± 0.91[c]	3.90 ± 0.23[ab]	0.46 ± 0.08[bc]
	TS-60 °C	10.00 ± 1.04[c]	3.86 ± 0.14[ab]	0.49 ± 0.05[abc]
Storage	Day 0	10.89 ± 0.21[a]	3.76 ± 0.03[c]	0.49 ± 0.04
	Day 14	9.59 ± 0.89[b]	3.86 ± 0.11[b]	0.47 ± 0.05
	Day 28	10.05 ± 0.72[b]	4.03 ± 0.17[a]	0.47 ± 0.07
	4 °C	10.65 ± 0.35[a]	3.82 ± 0.11	0.50 ± 0.03[a]
	25 °C	9.90 ± 1.02[b]	3.90 ± 0.18	0.46 ± 0.06[b]

[a-c] Means superscript with different alphabets in the same column showed the effect of treatments are significantly different (p<0.05). [a-c] Means superscript with different alphabets in the same column showed the effect of storage times and temperatures are significantly different (p<0.05).

It was found in the range of 3.77± 0.00–3.93 ± 0.12. Average pH values were affected by storage time and pH values increased as the storage time increased. Tiwari et al. [41] studied with US treated orange juice with a 20 kHz frequency, 40%, 70% and 100% amplitude and observed an increase in pH values after 30 days of storage at 10°C. It can be said that there is an increase in pH due to deterioration and fermentation during storage. During 28 days of storage, the effect of applied treatments on titration acidity was found to be decreased significantly (p<0.05) in TS-40°C and TS-50°C samples. Ordonez-Santos et al. [42] determined that there was no significant differences between the titration acidity values of fresh and US-treated fruit juices. It was reported that the energy levels at which sonication applications are performed are not sufficient to change the chemical structure of some large molecules. While the storage time had no effect on the titration acidity, the titratable acidity decreased as the storage temperature increased. The reduction in acidity is attributed to acidic hydrolysis of polysaccharides, in which acid is used to convert non-reducing sugars into reducing sugars [2]. Tomadoni et al. [43] found similar results after applying US (40 kHz, 10 and 30 minutes) to strawberry juice. It was reported that there was no significant difference in titration acidity compared to the control (p>0.05), and the changes after 10 days of storage at 5°C were not statistically significant (p>0.05).

Total Phenolics, Antioxidant Capacity, Total Carotenoids, Ascorbic Acid and HMF

The total phenolics, antioxidant capacity, total carotenoids, ascorbic acid, HMF amounts determined during the storage period of apple-carrot juice samples processed with various treatments are given in Table 5. Sonication application had a positive effect on the total phenolic substance and the US-treated samples presented higher phenolics levels (1400.00±146.11 to 1652.73±5.45 mg GAE/100 mL) when compared to the control (1472.73±10.91 mg GAE/100 mL). Also, it was determined that sonication treatments caused a higher increase in the total amount of phenolic content compared to the PAS treatment. Since many phenolic compounds are heat sensitive, it has been reported that the mild temperatures (40 and 60°C) used in TS treatment may cause less degradation of these compounds compared to the PAS treatment (90°C) [44]. The positive effect of US on phenolic substance content is attributed to the release of these compounds because of the damage to the cell wall after cavitation. In addition, it is stated that the amount of phenolic increases with the removal of oxygen from the environment by US [24]. During 28 days of storage, no significant change was observed in the total phenolic content of apple-carrot juices compared to the control (p>0.05).

Similar findings were observed in pear juice [20], gooseberry juice [42], pomelo juice and tangerine juice [44]. As the storage time increased (from day 0 to day 28), a decrease was observed in the total amount of phenolic substances. After 28 days of storage at 25°C, a significant (p<0.05) decrease was observed in the phenolic content compared to the samples stored at 4°C. Nadeem et al. [26] studied with US-treated carrot-grape juice on the total phenolic content during the storage period (1-90 days), the total phenolic substance amount decreased during the storage period. It has been reported that better results were obtained compared to the control sample. In another study, it was reported that US-treated carrot juice samples decreased gradually over 18 days of storage at 4°C [19]. The decrease in total phenolic content may occur as a result of oxidative destruction of some phenolic compounds or polymerization with proteins [39].

It was observed that an increase in antioxidant capacity values in the treated samples compared to the control. The effect of the applied treatments on the antioxidant capacity during storage was not significant (p>0.05)

Table 5. Total phenoliccontent, antioxidant capacity, total carotenoids, ascorbic acid and HMF properties of apple-carrot juice samples processed with various treatments during storage

		Total phenolic content (mg GAE/100 mL)	Antioxidant capacity (µmol trolox/mL)	Total carotenoids (mg/L β-carotene)	Ascorbic acid (mg/100 mL)	HMF (mg/L)
Treatments	Fresh	1472.73 ± 10.91[ab]	22.92 ± 1.13[b]	4.43 ± 0.07[ab]	48.55 ± 0.87[ab]	0.12 ± 0.01[c]
	PAS	1339.70 ± 196.15[b]	23.38 ± 2.02[b]	3.85 ± 0.47[c]	36.58 ± 4.29[d]	0.20 ± 0.06[a]
	US	1652.73 ± 5.45[a]	23.78 ± 0.74[ab]	4.84 ± 0.19[a]	49.71 ± 0.29[a]	0.14 ± 0.00[bc]
	TS-40°C	1594.91 ± 47.00[a]	23.75 ± 1.18[ab]	4.44 ± 0.37[ab]	43.84 ± 3.74[bc]	0.16 ± 0.02[abc]
	TS-50°C	1400.00 ± 146.11[b]	23.26 ± 1.18[b]	4.23 ± 0.21[b]	42.2 ± 5.21[c]	0.18 ± 0.05[ab]
	TS-60°C	1525.45 ± 161.98[a]	24.69 ± 1.21[a]	4.51 ± 0.57[ab]	41.04 ± 4.95[c]	0.19 ± 0.05[a]
Storage	Day 0	1592.73 ± 65.91[a]	24.23 ± 1.15[a]	4.18 ± 0.4[b]	47.34 ± 2.72[a]	0.14 ± 0.01[c]
	Day 14	1413.86 ± 189.70[b]	22.92 ± 1.63[b]	4.29 ± 0.51[ab]	38.55 ± 3.51[b]	0.18 ± 0.02[b]
	Day 28	1341.82 ± 151.78[b]	23.73 ± 1.49[ab]	4.54 ± 0.54[a]	36.88 ± 3.54[b]	0.22 ± 0.06[a]
	4°C	1522.60 ± 144.06[a]	24.11 ± 1.35[a]	4.44 ± 0.46[a]	43.25 ± 4.37	0.16 ± 0.02[b]
	25°C	1423.08 ± 188.99[b]	23.27 ± 1.53[b]	4.16 ± 0.48[b]	40.70 ± 6.70	0.20 ± 0.06[a]

[a-c] Means superscript with different alphabets in the same column showed the effect of treatments are significantly different (p<0.05). [a-c] Means superscript with different alphabets in the same column showed the effect of storage times and temperatures are significantly different (p<0.05).

compared to the control, but the change in TS-60°C was found to be significant (p<0.05). US-treated samples presented higher antioxidant capacity levels (23.26±1.18 to 24.69±1.21 μmol troloks/mL) when compared to the control samples (22.92 ± 1.13 μmol troloks/mL). The antioxidant capacity value increased as a result of the increase in the extraction of compounds with antioxidant characteristics, including phenolics, ascorbic acid, carotenoids etc. with the effect of cavitation in US process and the inactivation of some oxidation-related enzymes such as PPO enzymes due to the shear forces caused by sonication [45]. In addition, the phenolic compounds increase with the removal of active oxygen in the fruit juice by US treatment. And with this increase, antioxidant capacity may increase as a result of hydroxylation of hydroxyl radicals formed after US with the aromatic ring of polyphenols [18]. There were changes in the antioxidant capacity during 28 days of storage. Especially, samples at 14 days of storage showed more significant decrease (p<0.05). Similar decrease in the antioxidant capacity of grape-carrot juice processed with US (20 kHz frequency, 70% amplitude, 525 W, 5 minutes) during 90 days of storage was reported by Nadeem et al. [26] In another study, no significant change was observed in DPPH free radical scavenging activity and FRAP antioxidant capacity in the first week of storage at 4°C. However, these values were decreased after 21 days of storage and this decrease was because of the loss of antioxidants such as vitamin C, flavonoids and phenolics in fruit juices [39].

In current study, the average carotenoids ranged from 4.23±0.21 to 4.84±0.19 mg/L β-carotene, with no significant difference (p>0.05) between the US-treated samples and the control (4.43±0.07 mg/L β-carotene), for the treatment at US (alone), which presented higher values (4.84±0.19 mg/L β-carotene). It was found that PAS caused a significant decrease in the total carotenoid content (p<0.05) during storage. The increase in carotenoids in fruit juices exposed to US treatment is explained by the inactivation of the LOX enzyme and the release of these compounds as a result of cavitation [18, 19]. Thermal process can induce geometric isomerization of carotenoids and double bonds. It can cause the instability of the conjugated double bond of carotenoids and cause the loss of these compounds by oxidation [46]. Martinez-Flores et al. [37] also reported that the total carotenoid content increased by TS treatment performed at 50°C, 54°C and 58°C rate of 2.71%, 3.27% and 3.44%, respectively. In the current study, an increase in total carotenoids was detected as the storage time increased (p<0.05). However, the total carotenoid amounts of the samples stored at 4°C for 28 days were higher than the samples stored at 25°C (p<0.05). Similar increase in the total amount of carotenoid substance of carrot juice treated with sonication during 10 days of storage at 15°C was reported by Jabbar et al. [19].

In the current study, while an increase was observed in the ascorbic acid values of the samples treated with US alone (49.71±0.29 mg/100 mL), a decrease was detected in other treatments. Compared to the control, the ascorbic acid contents of PAS, TS-50°C, TS-60°C were degraded significantly ($p<0.05$) over the 28 days of storage at both 4°C and 25°C. The amount of ascorbic acid of the samples obtained by TS treatment decreased with an increasing temperature. Also, ascorbic acid amounts of apple-carrot juice samples treated with TS during storage were higher than the PAS-treated samples. Ascorbic acid in fruit juices is an effective quality indicator because it has antioxidant properties and is generally sensitive to processing conditions [44]. In some studies, it has been stated that hydroxyl radicals and other free radicals formed during US can reduce the amount of ascorbic acid between 6% and 36%, and in some other studies, US treatment was declared as effective in removing dissolved oxygen in the medium, thus preventing ascorbic acid oxidation and increasing its concentration in 2% to 34% [47]. Abid et al. [48] and Jabbar et al. [49] reported that TS treatment at 20°C increased the amount of ascorbic acid of juices. This effect is related to the removal of active oxygen from the fruit juice and prevent the stability of ascorbic acid. In the same studies, decreases in the amount of ascorbic acid were detected in treatments at 40 and 60°C, and it was stated that this effect could be caused by the severe physical conditions created by cavitation and collapsing bubbles. In the current study, the loss of ascorbic acid was higher in the samples stored at 25°C. Storage temperature can be effective in the degradation of ascorbic acid. Changes in the amount of ascorbic acid of the samples during storage, especially in the US treatment, it is connected to oxidation reactions by filling the cavitation bubbles with gases dissolved in fruit juice such as water vapor, O_2 and N_2 and interacting with free radicals. Oxidation can cause the conversion of ascorbic acid to dehydroascorbic acid during storage and decrease in its amount [50].

HMF, a product of Maillard reactions and ascorbic acid degradation, is significantly affected by processing and storage conditions. HMF increases significantly with the higher temperature, longer processing and storage time [51]. In the current study, an increase was observed in the HMF values of treated samples compared to the control, and this increase was significant ($p<0.05$) in PAS, TS-50°C, and TS-60°C samples. As the storage time and temperature increased, the HMF values also increased. Similar to the present work, Liao et al. [52] figured out that an increase in the HMF values of the clear red pitaya fruit juice exposed to thermal, US and TS treatments for 20 minutes or more. While this increase was statistically significant ($p<0.05$) in thermal and TS samples, it was determined that the increase in US samples was insignificant ($p>0.05$). Illera et al. [53] observed a significant increase in HMF values (0.75–1.3 mg/L) of apple juice processed with the TS treatment (20 kHz, 62°C, 100% amplitude and 20 min). This increase was because of the presence of oxygen or temperature. Also, it was reported that the processed samples caused an increase in the amount of HMF after 7 days of storage. The HMF content (0.12±0.01–0.2±0.06 mg/L) of all apple-carrot juice samples analyzed in the current study were determined below the 5 mg/L which is accepted as an indication of an excessive heat load in fruit juices [30].

Color Attributes and Browning Index (L^*, a^*, b^*, ΔE)

Color attributes and browning index values of apple-carrot juice samples stored at different temperatures are given in Table 6. Browning index absorbans values of treated samples were found in the range of 0.13±0.01 (US)–0.43±0.35 (PAS). It was found that processing methods increased the average browning index values during 28 days of storage, but this increase was not significant ($p>0.05$). The highest browning index values were found in PAS while the lowest browning index values were observed for the US-treated apple-carrot juice samples. Santhirasegaram et al. [18] found that sonicated samples showed smallest increase from the control to thermal treated juice samples. It was exhibited that juice sample subjected to PAS process accelerated formation of brown pigments, consequently darkening the juice when compared to mild heating treatments. In general, the initial browning index values of the samples increased after 28 days of storage, while this increase was greater in the samples stored at 25°C ($p<0.05$). Cruz-Cansino et al. [50] found that the browning values of the TS-treated purple cactus pear increased after 7 days of storage at 4°C for 28 days. Tiwari et al. [41] investigated the effects of different US applications on the orange juice at 10°C for 30 days and significant differences in browning index were found during storage. It has been stated that this change may be due to the breakdown of carotenoids. It has also been reported that ascorbic acid degradation causes non-enzymatic browning [54]. Reactive carbonyl groups, which are considered as non-enzymatic browning precursors, cause ascorbic acid degradation [55]. Storage temperature also affects the browning. Furfural accumulation is effective in browning, and a storage temperature of 12°C is reported to be critical in furfural accumulation [56].

Table 6. Color attributes and browning index of apple-carrot juice samples processed with various applications during storage

		Browning index (Abs)	$L*$	$a*$	$b*$	ΔE
Treatments	Fresh	0.14 ± 0.01	39.73 ± 0.42[b]	15.23 ± 0.55[d]	36.75 ± 0.57[c]	-
	PAS	0.43 ± 0.35	35.92 ± 1.27[c]	20.32 ± 1.64[a]	42.86 ± 0.9[a]	8.98 ± 1.49[a]
	US	0.13 ± 0.01	41.86 ± 0.81[a]	16.28 ± 0.48[cd]	37.41 ± 0.37[c]	2.62 ± 0.52[e]
	TS-40°C	0.36 ± 0.24	39.96 ± 2.19[b]	17.11 ± 1.48[bc]	38.93 ± 0.65[d]	3.88 ± 0.82[d]
	TS-50°C	0.34 ± 0.28	40.59 ± 2.03[ab]	17.81 ± 2.38[b]	39.38 ± 0.88[c]	4.88 ± 1.02[c]
	TS-60°C	0.4 ± 0.3	41.29 ± 2.55[a]	18 ± 2.52[b]	40.61 ± 0.76[b]	6.14 ± 0.81[b]
Storage	Day 0	0.16 ± 0.03[b]	41.31 ± 2.36[a]	15.84 ± 1.35[c]	39.16 ± 2.03[b]	4.56 ± 1.8[c]
	Day 14	0.57 ± 0.37[a]	39.07 ± 2.5[b]	19.04 ± 1.17[b]	40.6 ± 1.77[a]	6.02 ± 2.05[b]
	Day 28	0.41 ± 0.2[a]	37.36 ± 1.91[c]	20.42 ± 1.67[a]	40.76 ± 1.67[a]	7.27 ± 2.25[a]
	4°C	0.21 ± 0.08[b]	40.03 ± 2.62	17.59 ± 2.04	40.35 ± 2.11[a]	5.61 ± 2.28
	25°C	0.49 ± 0.35[a]	39.18 ± 2.95	18.39 ± 2.71	39.62 ± 1.83[b]	5.97 ± 2.32

[a-c] Means superscript with different alphabets in the same column showed the effect of treatments are significantly different (p<0.05). [a-c] Means superscript with different alphabets in the same column showed the effect of storage times and temperatures are significantly different (p<0.05).

All the colour parameters ($L*$ (lightness), $a*$ (redness) and $b*$ (yellowness)) were increased in the US treatment as compared to control during 28 days of storage. While the storage temperature did not affect (p>0.05) the average $L*$ and $a*$ values, it showed a significant (p<0.05) decrease in $b*$ values. While a decrease (p<0.05) was observed in the $L*$ value in the PAS sample, an increase was observed in US applications and this increase was found to be significant (p<0.05) in US and TS-60 °C (Table 6). While a significant increase in $L*$ values of thermosonicated carrot juice was observed, a decrease in $L*$ values of pasteurized carrot juice was observed has been reported by Jabbar et al. [49]. The increase may be due to the precipitation of unstable particles in the fruit juice by the TS process [57] and the increase in the turbidity value of the fruit juice with the homogenization resulting in the effect of sonication [14]. The $L*$ values decreased significantly over the time in all samples (p<0.05) and similar results were reported on strawberry juice [43] and carrot juice [36] during storage. The decrease in $L*$ values during storage is because of dark compounds formed as a result of non-enzymatic browning reaction [58, 59]. After the treatments during the 28 days storage period, an increase was observed in the $a*$ and $b*$ values of the PAS and TS samples. This increase was higher in the PAS samples in accordance with the higher temperatures. While the highest $a*$ and $b*$ values were detected on day 28, the lowest $a*$ and $b*$ values were detected on day 0 (for all samples). In current study, the $L*$ value of all samples decreased, while $a*$ and $b*$ values were increased at storage which shows the browning development as an action of time. Zia et al. [39] reported an increase in the $a*$ and $b*$ values of sugarcane juice processed with US treatment during 21 days of storage at 4°C. Cavitation causes carotenoid isomerization, partial precipitation of unstable suspended particles and degradation of compounds due to oxidation reactions, all of which cause changes in color parameters [60]. The highest total color difference (ΔE) was observed in PAS samples and the lowest total color difference was observed in US samples. Total color difference (ΔE) increased from 2.62±0.52 (US) to 8.98±1.49 (PAS) in accordance with the temperature. While the highest ΔE were detected on day 28, the lowest ΔE were detected on day 0 (for all samples). In US process, cavitation causes acceleration of chemical reactions, increase in diffusion rate, inactivation of enzymes and microorganisms and this affects the color change [61]. In thermal treatment, it is obtained darker appearance due to pigment destruction when the temperature increases, while high temperature accelerates the carotenoid isomerization. Therefore, the heat treatment application has a direct effect on the increase of ΔE values, and the higher the temperature, the higher the color change [16].

Sensory Analysis

Table 7 shows sensory analysis of freash and apple-carrot juice samples produced with PAS, US and TS applications. Color, appearance, smell, flavor and general taste scores.

Table 7. Sensory analysis of apple-carrot juice samples processed with various applications

Treatments	Sensory attributes				
	Color	Appearance	Odor	Flavor	Overall acceptability
Fresh	4.12 ± 0.76[a]	3.71 ± 0.67[abc]	3.59 ± 0.91[a]	3.76 ± 1.06[a]	3.94 ± 0.73[a]
PAS	2.82 ± 1.25[c]	2.12 ± 1.08[d]	2.82 ± 1.29[b]	2.53 ± 1.19[b]	2.65 ± 1.13[b]
US	3.18 ± 1.10[bc]	3.35 ± 0.84[c]	3.65 ± 0.97[a]	3.71 ± 0.82[a]	3.65 ± 0.97[a]
TS-40 °C	3.76 ± 0.81[ab]	3.65 ± 0.68[bc]	3.76 ± 0.81[a]	3.65 ± 1.19[a]	3.76 ± 0.88[a]
TS-50 °C	4.41 ± 0.84[a]	4.24 ± 0.73[a]	3.35 ± 0.97[a]	3.29 ± 0.75[a]	4.06 ± 0.80[a]
TS-60 °C	4.00 ± 0.97[a]	4.06 ± 0.64[ab]	3.59 ± 1.09[a]	3.65 ± 0.84[a]	3.88 ± 0.90[a]

[a-c] Means superscript with different alphabets in the same column differ significantly ($p<0.05$).

In current work, when the color and appearance values of apple-carrot juice samples were examined, the TS-50°C sample was most appreciated, while the PAS was least appreciated (p>0.05). TS-40°C (3.76 ± 0.81) sample was the most liked in terms of odor, while the least liked was PAS (2.82±1.29) sample. Pasteurization affected the odor values of fresh apple-carrot juice negatively (p<0.05), while the effect of sonication applications on odor was found to be statistically insignificant (p>0.05). The flavor scores of apple-carrot juice samples were found between 2.53±1.19 and 3.76±1.06. While the control sample was the most liked in terms of flavor, the least liked sample was PAS. However, no statistically significant difference was observed in terms of flavor between untreated and sonicated apple-carrot juice samples (p>0.05). When examined in terms of general taste, there was no statistically significant difference between the fresh apple-carrot juice sample and sonicated samples (p>0.05), while the most liked was determined as TS-50°C sample. The PAS sample was found as the least favored sample statistically (p<0.05).

Conclusion

In current study, a mixture of apple and carrot juice prepared at a ratio of 60:40 was processed with traditional PAS at 90°C for a min, US for 30 min and TS at 40°C, 50°C and 60°C for 30 min. The collected apple-carrot juice mixture was stored at both room temperature (25±0.5°C) and cold storage (4±0.5°C) for 28 days. The quality and microbiological analyses of control (untreated), PAS, US and TS-treated samples were performed on days 0, 14 and 28.

All treatments provided significant inactivation of aerobic mesophilic bacteria and yeast-molds compared to the control sample on day 0. During 28 days of storage, no bacteria and yeast-mold growth was observed in PAS, TS-50°C and TS-60°C samples, and it was determined that the shelf life of these samples continued and they were consumable after storage. Acidity, pH, brix, browning index and color that are quality characteristics were not changed with the use of sonication treatments. Total phenolics, antioxidant capacity, total carotenoids, higher retention of ascorbic acid and lower level of HMF were obtained with the sonication treatments compared to the heat treatments. Sonication and TS have a positive effect on color, appearance, odor, flavor and general taste compared to the PAS. Thus, considering all these aspects, TS at 50°C and 60°C could be alternative to heat treatment, further it can be successfully applied in the beverage industry for the production of apple-carrot juice.

Acknowledgment

This publication was produced from a master's thesis.

References

[1] Slavin, J. L., and Lloyd, B. (2012). Health benefits of fruits and vegetables. *Advances in Nutrition,* 3 (4), 506–516.
[2] Bhardwaj, R. L., and Pandey, S. (2011). Juice blends-a way of utilization of underutilized fruits, vegetables, and spices: a review. *Critical Reviews in Food Science and Nutrition,* 51, 563–570.
[3] Küçükkömürler, S., and Karakuş, S. Ş. (2009). Elma, Sağlık ve Kültür. *Tarım Bilimleri Araştırma Dergisi 2 (1), 183-186.*
[4] Sharma, K. D., Attri, S., Karki, S., and Thakur, N. S. (2012). Chemical composition functional properties and processing of carrot—a Review. *Food Science and Technology,* (49), 22–32.
[5] Gao, J., and Rupasinghe, H. P. V. (2012). Nutritional, Physicochemical and Microbial Quality of Ultrasound-Treated Apple-Carrot Juice Blends. *Food and Nutrition Sciences,* 3, 212-218.
[6] Anaya-Esparza, L. M., Velázquez-Estrada, R. M., Roig, A. X., García-Galindo, H. S., Sayago-Ayerdi, S. G. and Montalvo-González, E. (2017). Thermosonication: An alternative processing for fruit and vegetable juices. *Trends in Food Science & Technology,* 61, 26–37.
[7] Butz, P., and Tauscher, B. (2002). Emerging technologies: chemical aspects. *Food Research International,* 35(2–3), 279–284.
[8] Feng, H., Yang, W., and Hielscher, T. (2008). Power ultrasound. *Food Science and Technology International,* 14, 433-436.
[9] Mason, T. J., Paniwnyk, L., and Chemat, F. (2003). Ultrasound as a preservation technology, in: *Food Preservation Techniques,* 303–337.
[10] Piyasena, P., Mohareb, E., and McKellar, R. C. (2003). Inactivation of microbes using ultrasound: a review. *International Journal of Food Microbiology,* 87 (3), 207-216.
[11] Chemat, F., Zill-e-Huma, and Khan, M. K. (2011). Applications of ultrasound in food technology: Processing, preservation and extraction. *Ultrasonics Sonochemistry,* (18), 813-835.
[12] O'Donnell, C. P., Tiwari, B. K., Bourke, P., and Cullen, P. J. (2010). Effect of ultrasonic processing on food enzymes of industrial importance. *Trends in Food Science and Technology,* 21, 358-367.

[13] Aadil, R. M., Khalil, A. A., Rehman, A., Khalid, A., Inam-ur-Raheem, M., Karim, A., ... and Afraz, M. T. (2020). Assessing the impact of ultrasonication and thermo-ultrasound on antioxidant indices and polyphenolic profile of apple-grape juice blend. *Journal of Food Processing and Preservation, 44*, e14406.

[14] Tiwari, B. K., Muthukumarappan, K., O'Donnell, C. P., and Cullen, P. J. (2008). Effects of sonication on the kinetics of orange juice quality parameters. *Journal of Agricultural and Food Chemistry, 56*, 2423-2428.

[15] Adekunte, A. O., Tiwari, B. K., Cullen, P. J., Scannell, A. G. M., and O'Donnell, C. P. (2010). Effect of sonication on colour, ascorbic acid and yeast inactivation in tomato juice. *Food Chemistry, 122*, 500-507.

[16] Bermúdez-Aguirre, D., and Barbosa-C_anovas, G. V. (2012). Inactivation of Saccharomyces cerevisiae in pineapple, grape and cranberry juices under pulsed and continuous thermo-sonication treatments. *Journal of Food Engineering, 108*, 383-392.

[17] Abid, M., Jabbar, S., Wu, T., Hashim, M. M., Hu, B., Lei, S., ... and Zeng, X. (2013). Effect of ultrasound on different quality parameters of apple juice. *Ultrasonics Sonochemistry, 20*, 1182–1187.

[18] Santhirasegaram, V., Razali Z., and Somasundram C. (2013). Effects of Thermal Treatment and Sonication on Quality Attributes of Chokanan Mango (*Mangifera indica* L.) Juice. *Ultrasonics Sonochemistry, 20* (5), 1276-1282.

[19] Jabbar, S., Abid, M., Hu, B., Wu, T., Hashim, M. M., Lei, S., ... and Zeng, X. (2014). Quality of carrot juice as influenced by blanching and sonication treatments. *LWT - Food Science and Technology, 55*, 16–21.

[20] Saeeduddin, M., Abid, M., Jabbar, S., Wu, T., Hashim, M. M., Awad, F. N., Hu, B., Lei, S., and Zeng, X. (2015). Quality Assessment of Pear Juice Under Ultrasound and Commercial Pasteurization Processing Conditions. *LWT- Food Science and Technology, 64*, 452-458.

[21] Mohideen, F. W., Solval, K. M., Li, J., Zhang, J., Chouljenko, A., Chotiko, A., ... and Sathivel, S. (2015). Effect of continuous ultra-sonication on microbial counts and physico-chemical properties of blueberry (Vaccinium corymbosum) juice. *LWT - Food Science and Technology, 60* (1), 563–570.

[22] Fonteles, T. V., Costa, M. G. M., de Jesus, A. L. T., de Miranda, M. R. A., Gao, H. P., and Rupasinghe, V. (2012). Nutritional. Physicochemical and Microbial Quality of Ultrasound-Treated Apple-Carrot Juice Blends. *Food and Nutrition Sciences,* (3), 212-218.

[23] Aadil, R. M., Zeng, X.-A., Han, Z., and Sun, D. W. (2013). Effects of ultrasound treatments on quality of grapefruit juice. *Food Chemistry, 141*, 3201–3206.

[24] Bhat, R., Kamaruddin, N. S. B. C., Min-Tze L., and Karim A. A. (2011). Sonication improves kasturi lime (Citrus microcarpa) juice quality, *Ultrasonics Sonochemistry, 18*, 1295–1300.

[25] Yildiz, G., and Aadil, R.M. (2020). Comparison of high temperature-short time and sonication on selected parameters of strawberry juice during room temperature storage. *Journal of Food Science and Technology, 57* (4), 1462-1468.

[26] Nadeem, M., Ubaid, N., Qureshi, T. M., Munir, M., and Mehmood, A. (2018). Effect of ultrasound and chemical treatment on total phenol, flavonoids and antioxdant

properties on carrot-grape juice blend during storage. *Ultrasonics Sonochemistry*, 45, 16.

[27] Yildiz, G. (2019). Application of ultrasound and high-pressure homogenization against high temperature-short time in peach juice. *Journal of Food Process Engineering*, 42 (3), e12997.

[28] Yildiz, G., and Feng, H. (2019). Sonication of Cherry Juice: Comparison of Different Sonication Times on Color, Antioxidant Activity, Total Phenolic and Ascorbic Acid Content. *Latin American Applied Research Journal*, 49 (4), 255-260.

[29] AOAC. (2000). *Official methods of analysis*. 17th Edition, The Association of Official Analytical Chemists, Gaithersburg, MD, USA. Methods 925.10, 65.17, 974.24, 992.16

[30] Cemeroğlu, B. S. (2011). Meyve ve Sebze İşleme Teknolojisi, Ankara: *Nobel Yayınevi*, 2011.

[31] Cemeroğlu, B. (2004). Meyve ve sebze işleme teknolojisi 1. Cilt. *Gıda Teknolojisi Derneği Yayınları* No: 35, Ankara, 77-88.

[32] İzli, G. (2017). Total phenolics, antioxidant capacity, colour and drying characteristics of date fruit dried with different methods. *Food Science and Technoogyl, Campinas*, 37 (1), 139-147.

[33] Lee, H., Zhou, B., Liang, W., Feng, H., and Martin, S. E. (2009). Inactivation of Escherichia coli cells with sonication, manosonication, thermosonication, and manothermosonication: Microbial responses and kinetics modeling. *Journal of Food Engineering*, 93, 354-364.

[34] De Ritter, E., and Purcell, A. E. (1981). Carotenoid analytical methods. page 815 in *Carotenoids as Colorants and Vitamin A Precursors*. J. C. Bauernfeind, ed. Academic Press Inc., London, UK.

[35] Caminiti, I. M., Palgan, I., Muñoz, A., Noci, F., Whyte, P., Morgan, D. J., Cronin, D.A., and Lyng, J. G. (2012). The Effect of Ultraviolet Light on Microbial Inactivation and Quality Attributes of Apple Juice. *Food and Bioprocess Technology*, 5, 680–686.

[36] Adiamo, O. Q., Ghafoor, K., Al-Juhaimi, F., Babiker, E. E., and Ahmed, I. A. M. (2018). Thermosonication process for optimal functional properties in carrot juice containing orange peel and pulp extracts. *Food Chemistry*, 245, 79–88.

[37] Martínez-Flores, H. E., Garnica-Romo, M. G., Bermúdez-Aguirre, D., Pokhrel, P.R., and Barbosa-Cánovas, G.V. (2015). Physico-chemical parameters, bioactive compounds and microbial quality of thermo-sonicated carrot juice during storage. *Food Chem.* 172: 650–656. doi: 10.1016/j.foodchem.2014.09.072.

[38] Lee, H. S., Castle, W. S., and Coates, G. A. (2001). High-performance liquid chromatography for the characterization of carotenoids in the new nweet orange (Earlygold) grown in Florida, USA, *Journal of Chromatography A*, 913, 371-377.

[39] Zia, S., Khan, M. R., Zeng, X. A., Sehrish, R. M., Shabbir, M. A., and Aadil, R. M. (2019). Combined effect of microwave and ultrasonication treatments on the quality and stability of sugarcane juice during cold storage, *International Journal of Food Science and Technology*, 54, 2563–2569.

[40] Zhang, X., Zhang, M., S., Devahastin, S., and Guo, Z. (2019). Effect of Combined Ultrasonication and Modified Atmosphere Packaging on Storage Quality of Pakchoi (Brassica chinensis L.). *Food and Bioprocess Technology*, 12, 1573–1583.

[41] Tiwari, B. K., O'Donnell, C. P., Patras, A., Brunton, N. P., and Cullen, P. J. (2009). Stability of anthocyanins and ascorbic acid in sonicated strawberry juice during storage. *European Food Research Technology*, 228, 717-724.

[42] Ordóñez-Santos, L. E., Martínez-Girón, J., and Arias-Jaramillo, M. E. (2017). Effect of ultrasound treatment on visual color, vitamin C, total phenols, and carotenoids content in Cape gooseberry juice. *Food Chemistry*, 233, 96–100.

[43] Tomadoni, B., Cassani, L., Viacava, G., Moreira, M. D. R., and Ponce, A. (2017). Effect of ultrasound and storage time on quality attributes of strawberry juice. *Journal of Food Process Engineering*, 40 (5), 1–8.

[44] Alves, L. L., Santos, R. L., Bayer, B. L., Devens, A. L. M., Cichoski, A. J., and Mendonça, C. R. B. (2020). Thermosonication of tangerine juice: Effects on quality characteristics, bioactive compounds, and antioxidant activity. *Journal of Food Processing and Preservation*, 44, e14914.

[45] Wang, L., Kim, D., and Lee, C. (2000). Effects of heat processing and storage on flavanols and sensory qualities of green tea beverage. *Journal of Agricultural and Food Chemistry*, 48 (9), 4227–32.

[46] Rodriguez-Amaya, D. B. (2001). *A guide to carotenoid analysis in foods.* Washington, DC: ILSI Press.

[47] Aguilar, K., Garvín, A., Ibarz, A., and Augusto, P. E. D. (2017). Ascorbic acid stability in fruit juices during thermosonication. *Ultrasonics Sonochemistry*, 37, 375–381.

[48] Abid, M., Jabbar, S., Hu, B., Hashim, M. M., Wu, T., Lei, S., Khan, K. A., and Zeng, X. (2014). Thermosonication as a potential quality enhancement technique of apple juice. *Ultrasonics Sonochemistry*, 21, 984-990.

[49] Jabbar, S., Abid, M., Hu, B., Hashim, M. M., Lei, S., Wu, T., and Zeng, X. (2015). Exploring the potential of thermosonication in carrot juice processing. *Journal of Food Science and Technology*, 52, 7002-7013.

[50] Cruz-Cansino, N. D. S., Ramírez-Moreno, E., León-Rivera, J. E., Delgado-Olivares, L., Alanís-García, E., Ariza-Ortega, J. A., and Jaramillo-Bustos, D. P. (2015). Shelf life, physicochemical, microbiological and antioxidant properties of purple cactus pear (*Opuntia ficus* indica) juice after thermoultrasound treatment, *Ultrasonics Sonochemistry*, 27 (1), 277–286.

[51] Chaikham, P., Kemsawasd, V., and Apichartsrangkoon, A. (2016). Effects of conventional and ultrasound treatments on physicochemical properties and antioxidant capacity of floral honeys from Northernailand, *Food Bioscience*, 15, 19–26.

[52] Liao, H., Zhu, W., Zhong, and K., Liu, Y. (2020). Evaluation of colour stability of clear red pitaya juice treated by thermosonication. *LWT - Food Science and Technology*, 121, e108997.

[53] Illera, A. E., Beltrán, S., and Sanz, M. T. (2020). Enzyme inactivation and changes in the properties of cloudy apple juice after high-pressure carbon dioxide and

thermosonication treatments and during refrigerated storage. *Journal of Food Procesing And Preservation*, *44*, e14521.

[54] Marcy, J. E., Graumlich, T. R., Crandall, P. G., and Marshall, M. R. (1984). Factors affecting storage of orange concentrated. *Journal of Food Science*, *49*, 1628–1629.

[55] Joslyn, M. A. (1957). Physiological and enzymological aspects of juice production. In: *Fruit and Vegetable Juice* (edited by D. K. Tressler &M. A. Joslyn). Pp. *64*–115. Westport, CT: AVI Publishing.

[56] Kanner, J., Fishbein, J., Shalom, P., Harel, S., and Ben-Gare, I. (1982). Storage stability of orange juice concentrate packaged aseptically. *Journal of Food Science*, *47*, 429–436.

[57] Rawson, A., Tiwari, B. K., Patras, A., Brunton, N., Brennan, C., Cullen, and P. J., O'Donnell, C. (2011). Effect of Thermosonication on Bioactive Compounds in Watermelon Juice. *Food Research International*, (44), 1168–1173.

[58] Zhu, W., Ai, Y., Fang, F., and Liao, H. (2021). Application of Thermosonication in Red Pitaya Juice Processing: Impacts on Native Microbiota and Quality Properties during Storage. *Foods*, 10, 1041.

[59] Yildiz, G., İzli, G., and Aadil, R. M. (2020). Comparison of chemical, physical, and ultrasound treatments on the shelf life of fresh-cut quince fruit (*Cydonia oblonga* Mill.). *Journal of Food Processing and Preservation*, *44*, e14366.

[60] Khandpur, P., and Gogate, P. R. (2015). Effect of novel ultrasound-based processing on the nutrition quality of different fruit and vegetable juices. *Ultrasonics Sonochemistry*, *27*, 125–136.

[61] Sala, F. J., Burgos, J., Condon, S., Lopez, P., and Raso, J. (1995). Effect of heat and ultrasound on microorganisms and enzymes. In *New Methods of Food Preservation*; Gould, G. W.; Ed.; Blackie Academic and Professional Publisher: London, 177–203.

Chapter 4

Ultrasound-Assisted Extraction of Pectin from Tropical Fruit By-Products

Leticia Xochitl López-Martínez[1,*] Manuel Alejandro Vargas-Ortiz[2] and Ebber Addí Quintana-Obregón[1]

[1]CONACYT-Centro de Investigación en Alimentación y Desarrollo A.C. Sonora, México Coordinación de Alimentos de Origen Vegetal, Hermosillo, Sonora, México
[2]CONACYT-CIAD (Centro de Investigación en Alimentación y Desarrollo), Laboratorio de calidad, autenticidad y trazabilidad de los alimentos, Hermosillo, Sonora, México

Abstract

Pectin is a structural heteropolysaccharide ubiquitously found in the cell wall of higher plants. It is conventionally used to prepare jam, jellies, and marmalades in addition to being used in the food industry as a stabilizer and fat substitute. Besides, pectin exhibits bioactivities, such as satiety and decreasing glucose and lipids in the blood.

New applications of pectin continue to emerge; commercial pectin is extracted from apple pomace and citrus peels around the world. Tropical fruits and by-products could represent potential sources of pectin. Conventional extraction of pectin involves mineral acids and requires high energy and time consumption. Therefore, in line with caring for the environment, besides avoiding food waste, studies have been conducted to evaluate the efficiency of unconventional green extraction methods, such as ultrasound-assisted extraction of pectin from tropical fruits and their by-products. The present chapter discusses the potential applications of pectins and the use of ultrasound to obtain pectin from tropical by-products.

---

* Corresponding Author's Email: leticia.lopez@ciad.mx.

In: Power Ultrasound and Its Applications in Food Processing
Editors: Gulcin Yildiz and Gökçen Yıldız
ISBN: 979-8-88697-639-7
© 2023 Nova Science Publishers, Inc.

Keywords: by-products, tropical fruits, pectin, ultrasound

Introduction

The by-product generated during the fruit and vegetable industry processing is between 25 to 30% of the agricultural product loss [1]. Tropical fruit such as papaya, pineapple, mangosteen, banana, dragon fruit, and durian are processed to produce value-added products, including juices, jams, and dessert. The by-product includes seed, peel, pomace, and rind, that possess valuable bioactive molecules, mainly phenolic compounds, carotenoids, essential oils, pigments [2], and other secondary metabolites entrapped in tissue such as non-starch polysaccharides like pectin [3] a macromolecule composed of at least 17 different monosaccharides with over 20 types of different linkages [4]. Pectin extracted from tropical fruits has gained interest due to its particular bioactivities, such as antioxidant [5], anti-inflammatory [6], immunomodulatory [7], antidiabetic [6], modulation of gut microflora [8] and anticancer [9]. Moreover, the use of tropical fruit by-products to produce value-added products contributes to recycling waste, besides the recovery of pectin from fruit by-products could represent a strategy for the development of natural polymers [10].

The recovery of pectin involves extraction as a critical step; among the extraction methods, ultrasound-assisted extraction (UAE) shows advantages such as extraction at low temperature, less time and energy requirement and retention of the bioactive or functional characteristic of the pectin extract compared to conventional extraction. UAE extracts using high-intensity sound waves disrupt the plant tissue due to physical forces created during acoustic cavitation and help release extractable components in the solvent by enhancing mass transport [11]. The individual and interactive effects of the associated variables with UAE, such as frequency, power, duty cycle, temperature, time, solvent type, and solid-liquid ratio. Several researchers have studied the extraction of bioactive compounds using UAE from the fruit by-products [12, 13]; however, their study on the extraction of pectin from the by-products of tropical fruits is not very extensive. The present chapter discusses the potential applications of pectins and the use of ultrasound to obtain pectin from tropical by-products.

Chemical Structure

Up to about 30% of the cell wall consists of pectin, a complex heteropolysaccharide identified in plants' cell walls and middle lamellae [14]. Pectin is highly rich in galacturonic acid (GalA), which integrates the backbone of three more important domains found along with pectin: homogalacturonan (HGA) is a linear chain of 1,4-linked a-D-galacturonic acid (a-D-GalA) residues that are partially methyl-esterified at C-6 carboxyl groups [15]. HGA generally constitutes, on average, 65% of pectin. Rhamnogalacturonan-I (RG-I) region is highly branched, and 20 to 80% of the rhamnosyl residues and d-galacturonosyl residues [16]. The rhamnosyl residues are replaced by neutral sugar side chains, such as arabinan, arabinogalactan, and galactan [17], RGI, which constitutes 20 to 35. Rhamnogalacturonan-II (RG-II) is one of the highly conserved and complex structures which consist of distinct regions within HG, which make up about 10% of the pectin [16]. RG-II has side chains of sugar residue like 3-deoxy-manno-2-octulosonic, aceric acid, apiose-3-deoxy-lyxo-2-heptulosaric acid. The HG residues, alongside nine residues of GalA are bound to these side chains [14].

Source of Pectins

Commercial pectin is mainly derived from citrus peels, apple pomace and sugar beet pulp [18]. However, other sources, such as tropical fruit by-products, could represent an important source of pectin (Table 1), which clearly show the potential of this by-product for pectin obtention.

Table 1. Pectin content of some tropical fruits by-products

Tropical fruit	Part	Weight by-product (%)	Pectin in a by-product (%)	Yield (%)	Ref.
Banana	peel	20-30	4-6	2-3	[19]
Custard apple	peel	NR	NR	8.89	[20]
Dragon fruit	peel	5–10	17	2-17	[21]
Durian	rinds	50–60	NR	2–10	[22]
Fig	seeds	60–80	NR	5-6	[23]
Gac	pulp	NR	NR	53.80	[24]
Jackfruit	peel	NR	NR	14.5	[25]
Mango	peel	15–20	10–15	5–8	[26]

Table 1. (Continued)

Tropical fruit	Part	Weight by-product (%)	Pectin in a by-product (%)	Yield (%)	Ref.
Mangosteen	rind	10–30	NR	12	[27]
Papaya	peel	20–25	NR	11–50	[9]
Papaya	peel	NR	NR	25.4	[28]
Passion fruit	peel	50–60	15–20	8–10	[29]
Pineapple	peel	NR	NR	16.24	[30]

Ref.: References. NR. Not reported.

The extraction yield and the physicochemical characteristics of the recovered pectin depend on the extraction method and the solid-liquid ratio, temperature, time, pH and acid used during the extraction.

Extraction of Pectin

Several methods, including conventional hot water extraction methods and advanced or non-conventional techniques such as microwave-assisted extraction, enzyme-assisted extraction, pressurized liquid extraction, and ultrasound-assisted extraction, have been employed to extract pectin from natural sources. In addition, extraction methods for obtaining bioactive compounds, including pectin, have changed from a chemical point of view to an ecological point of view called "green chemistry" for sustainability [31].

Conventional hot water extraction of pectin requires a long extraction time, high energy expenditure, and sulfuric, nitric, and hydrochloric acid, which contradicts the principles of green chemistry.

The content of pectin in peel, skin, rinds, or seeds of tropical fruits has given rise to taking advantage of these residues to obtain pectin. The peel of fruits such as passion fruit (*Passiflora edulis*) gac (*Momordica cochinchinensis*), banana (*Musa paradisiaca*), mango (*Mangifera indica*), pineapple (*ananas comousus*), jackfruit (*Artocarpus heterophyllus*), and durian (*Durio*) have been used to obtain pectin for non-conventional extraction technique ultrasound-extraction (UAE) [5, 20, 21, 23, 24, 30].

Ultrasonic bath and ultrasonic probe units are the two modes of the UAE system. Bath is generally used; however, this system possesses disadvantages that could limit the reproducibility of the extraction, including the decrease of power with time; this causes the energy supplied to the bath is most of the time dissipated; another disadvantage is a deficiency in the uniformity of the ultrasound energy distribution. On the other hand, in the ultrasonic probes, the energy can be concentrated on a specific place of the sample generating higher efficiency in liquid cavitation [32]. In pectin extraction, probe units appear to be more used in research related to tropical fruit by-products (Table 2). Most of the research shown in Table 2 used statistical experimental designs to find the optimal extraction parameters (extraction and temperature time, solid-liquid ratio, US power, pH) to reach the maximum yield. Although strong mineral acids such as HCl, HNO$_3$ are cheaper, the use of organic acids particularly citric acid in combination with UAE, contributes to the development of "pectin produced green" or "clean labeled pectins" [33].

Table 2. Yield and conditions of UAE for the obtention of pectin from some tropical fruits by-products

Tropical fruit	Equipment	Conditions of extraction	Yield (%)	Ref.
White dragon fruit peel	Ultrasonic (GT SONIC-D6, China)	30 min at 75°C	16.30	[5]
Dragon fruit peel	Ultrasonic bath (Elmasonic P120H, Elma, Germany)	32 min at 85°C with a water-solid ratio 1:30 g/mL, pH 2.0 with citric acid. At 330 W and 37 kHz	23.09	[21]
	Ultrasonic bath (Powersonic P230D/HF) at 132 kHz and 80 W	70 min at 65°C, with a water-solid ratio 1:12 g/mL, pH 2		[34]
	(VCX750 Vibracell; USA), 40 kHz	15 min at 75°C	19.31	[35]
Gac pulp	Ultrasonic bath (Soniclean Pty Ltd., Thebarton, Australia) 50 Hz, and 200 W	90°C for 35 min with solvent to sample ratio of 50 mL/g and pH 1.5.	53.80	[24]
Pineapple peel	Ultrasonic (VCX 130, Sonics Vibra Cell, USA).	22 min a 70°C with a water-solid ratio 1:15.2 mL/g, pH 1.0, 20 kHz, 75% amplitude	16.24	[30]
Custard apple peel	Ultrasonic (VCX 130, Sonics Vibra Cell, USA). Probe: 2 cm flat tip	18.04 min at, 63.22°C with a water-solid ratio 1:23.5 g/mL, pH 2.3 with HCl and 20 Hz, Amplitude: 70%	8.93	[20]
Jackfruit peel	Ultrasonic (VCX 130, Sonics Vibra Cell, USA) Probe: 2 cm flat tip	24 min at 60°C, water-solid ratio: 5:1 mL/g, pH1.6, with citric acid and 20 Hz	14.5	[25]
Passion fruit peel	Ultrasonic (VCX 750, Newton, USA) Probe: 2 cm flat tip	20 min at 50°C, water-solid ratio: 1:30 mL/g, pH 2.0, with (1.0 M HNO$_3$, 20 Hz, 664 W	NR	[36]

Ref.: References. NR: Not reported.

It was observed that the yield of pectin from by-products of banana jam was enhanced by increasing ultrasound power from 100 to 350 W due to enlarged pores of the cell wall; however, above 350 W, the yield decreased, possibly due to the reduction of transmission of the ultrasound energy into the solvent. An increase in pectin yield was noted at pH 1.0 to 3.5 for an acidic solvent that hydrolyzes pectin's insoluble form into pectin's soluble form by reducing the viscosity, whereas, at values higher than pH 3.5, the pectin release was diminished. The yield was improved with sonication times of up to 28 min; however, more extended periods could cause structural modification and breakdown of the pectin. Varying the solid-liquid ratio from 1:5 to 1:21 g/mL increased the extraction effectiveness of pectin due to the expanded contact area between solid and solvent. Nevertheless, with ratios superior to 1:21 g/mL, the pectin yield was diminished due to the high viscosity made it difficult to generate cavitation [28].

In a study on the pectin extracted from dragon fruit peel, the authors observed that pectin yield increased from 5.52 to 6.84% when the sonication time was raised from 15 min to 35 min [21]. According to Zaid et al., [34], the main factors influencing the dragon fruit pectin yield are temperature, time, and pH. Moorthy et al., [37] isolated pectin from pomegranate peels. They observed that the higher dissolution of pectin in the solvent produces a higher yield. At pH 1.6, pectin yield was maximum, while pH higher than 1.6 may produce pectin aggregation hindering its release. The yield also improved by increasing the time up to 30 min, so it diminished at longer times due to the damage of pectin for structural decomposition. The rising temperature from 50 to 65°C improved the yield but diminished with the temperature increase due to pectin degradation. Moorthy et al., [25] obtained pectin from jack fruit peel, evaluating various extraction conditions to reach the highest yield. Ratio up to 18:1 mL/g enhanced the contact area between solvent and substrate, improving pectin release. At pH 1.0–1.7, higher pH values decreased the extraction yield. The viscosity was low, and then the yield improved. The authors noticed that improving the liquid-solid ratio up to 18:1 mL/g expanded the contact area between solvent and solid, improving pectin release. Sonication times up to 26 min enhanced the pectin yield due to the swelling of the solid pores supporting the solvent entrance and the pectin release. Greater extraction times produced structural breakdown and, therefore, decreased pectin yields. Temperatures up to 65°C improved the pectin yield due to loosening and swelling, improving solvent access to the material. Wang et al., [38] utilized mango peel for extracting pectin using citric acid as an extraction medium. UAE has a short process (15 min) and medium yields at 80°C (17.15%).

Biological Activities of Pectin

Pectin from tropical fruits by-products has various beneficial effects on human health due to its antioxidant, antidiabetic, anti-inflammatory, immunomodulatory, antifatigue, anticancer, healing wounds and reducing blood cholesterol levels. These biological activities are associated with their structures [39, 6, 40, 41, 9].

Pectin might act as a proton donor and could scavenge DPPH˙ and ABTS˙+ free radicals. Wang et al., [38] describe that the antioxidant capacity of pectin isolated from jackfruit peel is highly correlated with the content of monosaccharides capable of reducing compounds that supply proton, which reacts with radicals to finish the oxidation reaction generating stable products. Wathoni et al., [39] reported that the antioxidant activity in pectin from mangosteen rind was moderate $IC_{50} = 161.93$ μg/g compared with the control (ascorbic acid $IC_{50} = 3.39$ μg/g). The authors concluded that the antioxidant activity of the pectin depends on its structure and that the viscosity is not too high. On the other hand, the antioxidant activity of the extracted pectin from fig skin had a greater capacity to scavenge ABTS˙+ compared to DPPH; however, the reducing capacity of BHT was substantially superior to the extracted pectin. The antioxidant capacity of pectin can be attributed to the high number of hydroxyl and carboxyl groups in the pectin structure, particularly in GalA, as electron donors [42]. Xu et al., [43] studied the antioxidant activity of jackfruit peel. Pectin demonstrated considerable scavenging activity on DPPH˙ radicals in a dose-dependent way, while vitamin C showed a greater free radical-scavenging activity than pectin at identical concentrations. The value of pectin was $IC_{50} = 1.1$ mg/mL, while the ascorbic acid value was 0.5 mg/mL. The authors concluded that the antioxidant activities of the pectin were strongly correlated with their monosaccharide composition.

Silva et al., [6] estimated the effects of pectin from passion fruit on male Wistar rats with alloxan-induced diabetes were divided into diabetic control (treated with distilled water) and diabetic groups treated with pectin (2, 10, and 25 mg/kg) every day for 5 days orally administered. Glibenclamide and metformin were used as reference drugs that decreased 80 and 79% glucose concentrations. Forty-eight hours after the alloxan treatment and 5 days later, blood glucose concentration did not vary in the diabetic control group. The diabetic groups treated with pectin showed a decrease in blood glucose concentrations from 46 to 65% compared with the same group before the treatment that started 48 hours after alloxan administration.

Pectin from fruits has been associated with a beneficial effect on the human intestinal microflora. Incubation of the feces samples of healthy individuals in a liquid nutrient medium with apple pectin results in increased bifidobacteria content compared to the pectin-free medium. After eight healthy volunteers ate two apples per day for two weeks, the concentration of bifidobacteria in feces was raised on the 7 and 14 days. The amount of lactobacteria also increased. In the same time, the number of *Clostridium perfringens* significantly decreased, and the tendency to reduce pseudomonas was noted [40].

Klosterhoff et al., [41] studied the *in vivo* antifatigue activities of HG-rich pectin (52.1% galacturonic acid) from acerola (*Malpighia emarginata*). The authors reported that oral administration in albino rats for 28 days, at doses of 50, 100 and 200 mg/kg caused lengthened swimming time, where the animals showed increased times to exhaustion of 95, 151 and 129 min, respectively) compared to the control (53 min). Do Prado et al., [9] demonstrated that pectin of papaya pulp significantly decreased cell viability and induced necroptosis in prostate cancer cell line (PC3) and colon cancer cells (HCT116 and HT29). These activities are dependent on the presence of arabinogalactan type II (AGII) structure, which can interrupt the interaction between extracellular matrix proteins and cancer cells, enhancing cell detachment and promoting apoptosis or necroptosis.

The pectic polysaccharide PLE-II isolated from persimmon leaves showed antitumor and antimetastatic activities. The polysaccharides amplified natural killer (NK) cells, which caused cytotoxicity to lymphoma tumor cells. The authors concluded that NK cells possess a key role in controlling the antimetastatic effect of PLE-II [44].

Duan et al., [45] reported that a pectic polysaccharide from caqui (*Diospyros kaki*) leaves has GalA in their backbone could stimulate the lipopolysaccharide-induced B-lymphocyte proliferation.

Final Remarks

Tropical fruits are an excellent source of bioactive compounds, including pectins, that have been shown several bioactivities such as antioxidant, anti-inflammatory, immunomodulatory, antidiabetic, and antifatigue. Nowadays, consumer demand for obtaining natural ingredients such as pectin in an ecological approach that does not involve the use of chemical solvents has generated the need to use sustainable extraction methods. Ultrasound-assisted extraction is an environmentally friendly technique that enhances extraction efficiency, reduces energy consumption, and lowers the extraction time. In addition, ultrasound extraction has been useful in separating pectin from tropical fruit by-products.

References

[1] Sagar N. A., Pareek S., Sharma S., Yahia E. M., Lobo. M. G. Fruit and vegetable waste: Bioactive compounds, their extraction, and possible utilization. *Comprehensive Reviews in Food Science and Food Safety* 2018; 17:512-531.

[2] González-Aguilar G. A., Robles-Sánchez R. M., Martínez-Téllez M. A., Olivas G. I., Alvarez-Parrilla E., De La Rosa, L. A. Bioactive compounds in fruits: health benefits and effect of storage conditions. *Stewart Postharvest Review* 2008; 4:1-10.

[3] Enríquez-Valencia S. A., Gonzalez-Aguilar G. A., López-Martínez L. X. Tropical fruits and by-products as a potential source of bioactive polysaccharides. *Biotecnia* 2021; 23:125-132.

[4] Voragen A. G., Coenen G. J., Verhoef R. P., Schols H. A. Pectin, versatile polysaccharide in plant cell walls. *Structural Chemistry* 2009; 20:263-275.

[5] Nguyen B. M. N., Pirak T. Physicochemical properties and antioxidant activities of white dragon fruit peel pectin extracted with conventional and ultrasound-assisted extraction. *Cogent Food & Agriculture* 2019; 5:1633076.

[6] Silva D. C., Freitas A. L., Pessoa C. D., Paula R. C., Mesquita J. X., Leal L. K., Brito G. A., Gonçalves D. O., Viana, G. S. Pectin from *Passiflora edulis* shows anti-inflammatory action as well as hypoglycemic and hypotriglyceridemic properties in diabetic rats. *Journal of Medicinal Food* 2011; 14:1118-1126.

[7] Holderness J., Schepetkin I. A., Freedman B., Kirpotina L. N., Quinn M. T., Hedges J. F., Jutila M. A. Polysaccharides isolated from Acai fruit induce innate immune responses. *PloS one* 2011; 6: e17301.

[8] Blanco-Pérez F., Steigerwald H., Schülke S., Vieths S., Toda M., Scheurer S. The dietary fiber pectin: Health benefits and potential for the treatment of allergies by modulation of gut microbiota. *Current Allergy and Asthma Reports* 2021; 21: 1-19.

[9] Do Prado S. B. R. D., Ferreira G. F., Harazono Y., Shiga T. M., Raz A., Carpita N. C., Fabi J. P. Ripening-induced chemical modifications of papaya pectin inhibit cancer cell proliferation. *Scientific Reports* 2017; 7: 1-17.

[10] Picot-Allain M. C. N., Ramasawmy B., Emmambux M. N. Extraction, characterisation, and application of pectin from tropical and sub-tropical fruits: a review. *Food Reviews International* 2022; 38: 282-312.

[11] Zhu Z., Wu Q., Di X., Li S., Barba F. J., Koubaa M., Roohinejad S., Xiong X., He J. Multistage recovery process of seaweed pigments: Investigation of ultrasound assisted extraction and ultra-filtration performances. *Food and Bioproducts Processing* 2017; 104:40-47.

[12] Kumar K., Srivastav S., Sharanagat V. S. Ultrasound assisted extraction (UAE) of bioactive compounds from fruit and vegetable processing by-products: A review. *Ultrasonics Sonochemistry* 2021; 70:105325.

[13] Patra A., Abdullah S., Pradhan R. C. Review on the extraction of bioactive compounds and characterization of fruit industry by-products. *Bioresources and Bioprocessing* 2022; 9: 1-25.

[14] Cui J., Zhao C., Feng L., Han Y., Du H., Xiao H., Zheng, J. Pectins from fruits: Relationships between extraction methods, structural characteristics, and functional properties. *Trends in Food Science & Technology* 2021; 110: 39-54.

[15] Mao G., Wu D., Wei C., Tao W., Ye X., Linhardt R. J., Orfila C., Chen S. Reconsidering conventional and innovative methods for pectin extraction from fruit and vegetable waste: Targeting rhamnogalacturonan I. *Trends in Food Science and Technology* 2019; 94: 65-78.

[16] Caffall K. H., Mohnen D. The structure, function, and biosynthesis of plant cell wall pectic polysaccharides. *Carbohydrate Research* 2009; 344: 1879-1900.

[17] Romdhane M. H., Beltifa A., Mzoughi Z., Rihouey C., Mansour H. B., Majdoub H., Le Cerf D. Optimization of extraction with salicylic acid, rheological behavior and antiproliferative activity of pectin from Citrus sinensis peels. *International Journal of Biological Macromolecules* 2020; 159, 547-556.

[18] Li D., Wang L. J. Characterization of pectin extracted from sugar beet pulp under different drying conditions. *Journal of Food Engineering* 2017; 211, 1-6.

[19] Oliveira T. Í. S., Rosa M. F., Cavalcante F. L., Pereira P. H. F., Moates G. K., Wellner N., Mazzetto S. E., Waldron K. W., Azeredo H. M. Optimization of pectin extraction from banana peels with citric acid by using response surface methodology. *Food Chemistry* 2016; 198, 113-118.

[20] Shivamathi C. S., Moorthy I. G., Kumar R. V., Soosai M. R., Maran J. P., Kumar R. S., Varalakshmi P. Optimization of ultrasound assisted extraction of pectin from custard apple peel: Potential and new source. *Carbohydrate Polymers* 2019; 225, 115240.

[21] Chua B., Tang S. F., Ali A., Chow Y. H. Optimisation of pectin production from dragon fruit peels waste: drying, extraction and characterisation studies. *SN Applied Sciences* 2020; 2: 1-13.

[22] Maran J. P. Statistical optimization of aqueous extraction of pectin from waste durian rinds. *International journal of biological macromolecules* 2015; 73, 92-98.

[23] Liang R. H., Chen J., Liu W., Liu C. M., Yu W., Yuan M., Zhou X. Q. Extraction, characterization and spontaneous gel-forming property of pectin from creeping fig (Ficus pumila Linn.) seeds. *Carbohydrate Polymers* 2012; 87: 76-83.

[24] Tran T. T. B., Vu Q. L., Pristijono P., Kirkman T., Nguyen M. H., Vuong Q. V. Optimizing conditions for the development of a composite film from seaweed hydrocolloids and pectin derived from a fruit waste, gac pulp. *Journal of Food Processing and Preservation* 2021; 45: e15905.

[25] Moorthy I. G., Maran J. P., Ilakya S., Anitha S. L., Sabarima S. P., Priya B. Ultrasound assisted extraction of pectin from waste *Artocarpus heterophyllus* fruit peel. *Ultrasonics Sonochemistry* 2017; 34: 525-530.

[26] Sudhakar D. V., Maini S. B. Isolation and characterization of mango peel pectins. *Journal of Food Processing and Preservation* 2000; 24(3): 209-227.

[27] Wathoni N., Shan C. Y., Shan W. Y., Rostinawati T., Indradi R. B., Pratiwi R., Muchtaridi, M. Characterization and antioxidant activity of pectin from Indonesian mangosteen (Garcinia mangostana L.) rind. *Heliyon* 2019; 5: e02299.

[28] Maran J. P., Prakash K. A. Process variables influence on microwave assisted extraction of pectin from waste *Carcia papaya* L. peel. *International Journal of Biological Macromolecules* 2015; 73, 202-206.

[29] Freitas C. M. P., Sousa R. C. S., Dias M. M. S., Coimbra J. S. R. Extraction of pectin from passion fruit peel. *Food Engineering Reviews* 2020; 12(4): 460-472.

[30] Shivamathi C. S., Gunaseelan S., Soosai M. R., Vignesh N. S., Varalakshmi P., Kumar R. S., Karthikumar S., Kumar R. V., Baskar R., Rigby S. P., Syed A., Elgorban A. M., Moorthy, I. M. G. Process optimization and characterization of pectin derived from underexploited pineapple peel biowaste as a value-added product. *Food Hydrocolloids* 2022; 123: 107141.

[31] García A., Rodríguez-Juan E., Rodríguez-Gutiérrez G., Rios J. J., Fernández-Bolaños J. Extraction of phenolic compounds from virgin olive oil by deep eutectic solvents (DESs). *Food Chemistry* 2016; 197, 554-561.

[32] Luque-Garcia J. L., De Castro M. L. Ultrasound: a powerful tool for leaching. *TrAC Trends in Analytical Chemistry* 2003; 22(1), 41-47.

[33] Nadar C. G., Arora A., Shastri Y. *Sustainability Challenges and Opportunities in Pectin Extraction from Fruit Waste*, 2022, ACS Engineering Au.

[34] Zaid R. M., Mishra P., Noredyani A. S., Tabassum S., Ab Wahid Z., Sakinah A. M. Proximate characteristics and statistical optimization of ultrasound-assisted extraction of high-methoxyl-pectin from Hylocereus polyrhizus peels. *Food and Bioproducts Processing* 2020; 123: 134-149.

[35] Van M. P., Thi H. H. P. Physiochemical properties and chemical structures of dragon fruit peel pectin extracted by conventional and ultrasound assisted extraction methods. *SEATUC Journal of Science and Engineering*, 2020; 1: 39-43.

[36] de Oliveira C. F., Giordani D., Lutckemier R., Gurak P. D., Cladera-Olivera F., Marczak L. D. F. Extraction of pectin from passion fruit peel assisted by ultrasound. *LWT-Food Science and Technology* 2016; 71: 110-115.

[37] Moorthy I. G., Maran J. P., Muneeswari S., Naganyashree S., Shivamathi C. S. Response surface optimization of ultrasound-assisted extraction of pectin from pomegranate peel. *International Journal of Biological Macromolecules* 2015; 72: 1323-1328.

[38] Wang M., Huang B., Fan C., Zhao K., Hu H., Xu X., Pan S., Liu F. Characterization and functional properties of mango peel pectin extracted by ultrasound assisted citric acid. *International Journal of Biological Macromolecules* 2016; 91: 794-803.

[39] Wathoni N., Shan C. Y., Shan W. Y., Rostinawati T., Indradi R. B., Pratiwi R., Muchtaridi M. Characterization and antioxidant activity of pectin from Indonesian mangosteen (Garcinia mangostana L.) rind. *Heliyon* 2019; 5: e02299.

[40] Shinohara K., Ohashi, Y., Kawasumi K., Terada A., Fujisawa T. Effect of apple intake on fecal microbiota and metabolites in humans. *Anaerobe* 2010; 16: 510-515.

[41] Klosterhoff R. R., Bark J. M., Glänzel N. M., Iacomini M., Martinez G. R., Winnischofer S. M., Cordeiro L. M. Structure and intracellular antioxidant activity of pectic polysaccharide from acerola (Malpighia emarginata). *International journal of biological macromolecules* 2018; 106: 473-480.

[42] Gharibzahedi S. M. T., Smith B., Guo Y. Pectin extraction from common fig skin by different methods: The physicochemical, rheological, functional, and structural evaluations. *International journal of biological macromolecules* 2019; 136: 275-283.

[43] Xu S. Y., Liu J. P., Huang X., Du L. P., Shi F. L., Dong R., Huang X. T., Zheng K., Liu Y., Cheong K. L. Ultrasonic-microwave assisted extraction, characterization and biological activity of pectin from jackfruit peel. *LWT* 2018; 90:577-582.

[44] Park H. R., Hwang D., Hong H. D., Shin K. S. Antitumor and antimetastatic activities of pectic polysaccharides isolated from persimmon leaves mediated by enhanced natural killer cell activity. *Journal of Functional Foods* 2017; 37:460-466.

[45] Duan J., Wang X., Dong Q., Fang J. N., Li X. Structural features of a pectic arabinogalactan with immunological activity from the leaves of Diospyros kaki. *Carbohydrate Research* 2003; 338:1291-1297.

Chapter 5

Value Added Ultrasound Technology – Assuring Safety of Seafood

Vasantha Subramoniam Pramitha[1] and Parameswara Panicker Sreejith[2,*]

[1]Department of Aquatic Biology and Fisheries, University of Kerala, Thiruvananthapuram, India
[2]Department of Zoology, University of Kerala, Thiruvananthapuram, India
Advanced Centre for Regenerative Medicine and Stem cell in Cutaneous Research (AcREM-Stem), University of Kerala, Thiruvananthapuram, India

Abstract

The demand for high-quality innovative meals has prompted the expansion of a number of non-thermal ways to food preparation and one of which has proven to be extremely useful ultrasound technology. The seafood sector is progressively focusing on new product development and implementing innovative processing technologies, necessitating the adoption of more quick and efficient processing methods. Application of ultrasonic energy significantly improves the drying effect, especially at moderate temperatures, thus minimizing food quality loss and the efficacy of ultrasound decreases with the increase in the drying temperature. Ultrasound technology's involvement in seafood processing has swiftly evolved to enhance conventional technologies that are detrimental to the product quality. This technology becomes more powerful when used in combination with conventional processing techniques and ultrasound assented unit operation reduces the processing time also. The goal of this chapter is to identify the benefits of ultrasound

---

[*] Corresponding Author's Email: p.sreejith@gmail.com, psreejith@keralauniversity.ac.in, pramivs_vzm@yahoo.com.

In: Power Ultrasound and Its Applications in Food Processing
Editors: Gulcin Yildiz and Gökçen Yıldız
ISBN: 979-8-88697-639-7
© 2023 Nova Science Publishers, Inc.

technology in the seafood industry that have been identified as promising for future growth. This chapter not only discusses how ultrasonography impacts seafood, but it also discusses an overview on seafood's spoilage and the recent innovative techniques of processing and preservation which are related with the extension of shelf life of fish and fishery products.

Keywords: quality, seafood processing, ultrasound-assisted extraction, ultrasound assisted freezing

Introduction

Marine habitat stands as a reservoir of large-scale organisms and biomolecules that have various applications in the sea food industry. The ocean's bioresource mass produce up to 75% more food for our growing populations and accords a major portion of the world's food supply. Seafood industry is increasingly moving towards new product development and adopting innovative processing methods that allow better and more efficient products to produce high quality and effectively processed and microbiologically safe products for the consumers [1].

Deterioration of the product before it reaches to the consumer is not uncommon and may lead to significant waste of fish resources, necessitating more rapid and efficient processing approaches [2]. In India, sea food industry has emerged as one of the major export-oriented industry and exports ninety varieties of seafood products to various countries all over the world. Frozen products dominate the trade, and it contains frozen shrimp, frozen fish, cuttlefish and squids.

Major sources for the production of fish byproducts include fish offal and trash fishes. Fish flesh on an average constitute 15-20% protein and some species of fish contains very high amounts of body oil. Few species of fish like shark, cod are good sources of liver oil and fish processing industries turn out large quantities of fishery waste to convert high quality protein, fat, minerals to produce different fishery byproducts. Fish protein concentrate, pearl essence and fish skin leather are other type of byproducts generally processed out from fish and fish waste. High economic value byproducts include chitin and chitosan processed out of shrimp, crab and other crustacean shell wastes. Biochemical and pharmaceutical products like bile salts, insulin, glucosamine are byproducts of great significance [3].

The role of technology in seafood processing has evolved rapidly in response to support innovation, productivity, waste reduction, waste recovery, utilization, increase shelf-life, improve food safety, and facilitate exports. Several innovative processing technologies have recently emerged as a result and ultrasonic method is one of those rapidly emerging techniques that were devised to minimize processing, enhance quality and safeguard the safety of food products [4].

As the sea food industry in India started diversifying into the production of various value-added fishery products, packing of these products become essential. Processing technologies have been developed for preparing fish ready to cook and ready to serve and value-added products from many varieties of cheaper fish has less demand in the fresh fish market. The most important such products are battered and breaded shrimp, squid rings, stuffed squid, fish fingers, fish cutlet and fish momos. The thermoformed trays produced from food grade PVC and poly styrene are suitable for packaging value added fishery products. These thermoformed trays are usually sealed with polyester based film and are useful in self-service system also [5].

Packaging is an important part of production, storage and distribution of any product and it facilitates the easier and safer transport of the product and protects it from contamination and loss, damage or degradation. It also provides an opportunity to the buyer to identify the product and persuades the buyer to purchase the product. Packaging also ensuring the safe delivery of a product to the end consumer in sound condition at the minimum overall cost. Food packaging provides an external means of preservation of food during storage, transportation and distribution. It facilitates storage, effective chilling, internal and long distance transport, easy determination of quantities and display in whole sale and retail markets. Packaging materials protect the product from contamination or loss [6].

The conventional system of processing shrimps is to in plate freezers as blocks of 2 kg and 4 kg and the material for packing comprises LDPE/HDPE films, duplex carton and corrugated fibre board box. The problems with this packaging system are their low mechanical strength and tendency to get wet due to deposit of moisture. Shrimps are frozen as blocks of 2 kg each in duplex board cartons lined with low density polyethylene (LDPE) and 10 such cartons are packed in master cartons made of 5 ply or 7 ply corrugated fibre board box [7].

Consumers always demand the best products regarding quality and nutrition and the trend of choosing products without preservatives, excessive heat treatments, and additives are also gaining popularity. To fulfill the desires of consumers new technologies are introduced in every food industry. Ultrasound is an emerging processing and analytical technique in food processing, and it shows great results in the processing of different products regarding reduction in processing time, product purity, simplifying the complex procedure, and nonthermal processing [8].

Ultrasound is a fast, versatile, emerging and promising non-destructive green technology with a wide range of applications including seafood processing. The advantages of using ultrasound for sea food processing include more effective mixing and micro-mixing, mass transfer, reduced thermal and concentration gradients, increased production, elimination of process steps, reduced temperature, selective extraction, reduced equipment size and faster response to the process [6]. Therefore, the objective of this chapter is to briefly discuss the wide diversity of underutilized marine biomass that could be further explored in the food industry by the application of innovative and emerging technologies like ultrasound. The chapter includes an overview on seafoods spoilage and the recent innovative techniques of processing and preservation which are related with the extension of shelf life of fish and fishery products [5].

Most of the world is running out of energy sources and ultrasound will provide them the best alternative regarding processing assistance. Keeping these uses in view, current chapter addresses the application of possible ultrasound implementation in fish processing sectors. The challenges faced by the industry in applying ultrasound in the fish industry are also mentioned here.

Extraction Of Lipids, Carotenoids and Other Marine Compounds by Ultrasound

The marine territory is a phenomenal biodiverse constituting nearly half of the total universal biodiversity. Greater species diversity ensures a huge number of distinct natural compounds. Sustainable recovery of bioactive from renewable sources is a need of the hour. Ultrasonic irradiation is an authenticated technology to boost the efficiency of the extraction of bioactive compounds from marine resources. The technique works on the principle of sonophysical cavitation effect resulting in changes in the physicochemical properties of the molecule, which accelerates the extraction rate and achieves better yield exceptionally at greener conditions [9].

An Overview of the Application of Emerging Technologies in the Bio-Marine Food Sector

The bioresources required for food production are diminishing and new approaches are needed to feed the current and future global population [8]. Seafood products are highly perishable, owing to their high-water activity, close to neutral pH and high content of unsaturated lipids and non-protein nitrogenous compounds. Thus, such products require immediate processing and packaging to retain their safety and quality. At the same time, consumers prefer fresh, minimally processed seafood products that maintain their initial properties [7]. Food safety, nutrient loss during processing, and more time consumption is the issue for these types of industries. Ultrasound provides great opportunity to achieve goals regarding food safety, nutrition, and more fast processing and it is also very effective to use as an analytical technique. Thermal processing usually destroys much of the nutrition in the food which is not acceptable for consumers [10].

This is an emerging technology that is why not much adaptation is being in use by industries. Ultrasound waves are very flexible to choose from regarding their velocity, attenuation, frequency spectrum when it passes through a medium. Their use in seafood processing for different processes depends on the wavelength according to the process and need [3]. Ultrasound is considered as a more efficient and reliable method to use for cell disruption. These waves create shear force and break the cell wall and this phenomenon can be used to extract the oil from fish and other seafood [11].

Ultrasound waves are not only used as an alternative to the thermal process, it also can be used in place of washing. This process is also very interesting as it also reduced the amount of wastewater because more than half of the wastewater comes from the product washing [3]. The use of ultrasound is also increasing in the field of checking the fat-lean portions of the fish prior to catch and death. Water holding capacity is a big concern of the meat and fish industry and it is shown that the meat treated with ultrasound has more WHC and for a longer period of processing [10]. All these facts showed that ultrasound is non-toxic, reliable and nutrition-safe treatment for foodstuff, but still is not practiced for fish and other seafood products. As we know that the future concern of everyone will be nutrition and sustainable food so the old and toxic methods of processing can create headaches for both industry and consumers. During processing when we provide heat to the amine dense product it creates some time carcinogens which is a big concern of consumers.

So, it is possible to apply ultrasound in fish and other seafood processing operations to overcome the problems associated with conventional processing methods. Enzymes and micro-organisms responsible for the spoilage of fish are listed in Table 1.

Table 1. Enzymes and micro-organisms responsible for the spoilage of fish

Spoilage	Class	Effects
Enzymatic spoilage	Glycolytic enzymes	Lactic acid production resulting in pH drop
	Autolytic enzymes	Gradual production of hypoxanthine
	Cathepsins	Softening of tissue
	Chymotrypsin, trypsin, carboxy-peptidases	Belly-bursting
	Calpain	Softening
	Collagenases	Softening and gaping of tissue
	Trimethylamine Oxide (TMAO) demethylase	Formaldehyde production
Microbial spoilage	Bacterial species	Pseudomonas, Alcaligenes, Vibrio, Serratia and Micrococcus
	Gram-negative fermentative bacteria	Vibrionaceae
	Psychrotolerant gramnegative bacteria	Pseudomonas spp. and Shewanella spp
	Pathogenic bacteria	Streptococcus iniae, Vibrio vulnificus, Salmonella spp, Clostridium botulinum type E, Erysipelothrix rhusiopathiae

Ultrasound Processing

Ultrasound processing refers to the use of high frequency sound waves (>20 MHz) which cause physical, mechanical or chemical changes in the food material. The two types of sound waves which have huge potential in food processing applications include high frequency low energy power ultrasound and low frequency high energy power ultrasound. Low energy ultrasound has frequency in the range of 100 KHz with intensities less than 1W/cm2 and used for monitoring of food products during processing and storage. Its other potential applications are evaluation of the composition of meat products, fish of raw and fermented stages and checking quality of fruits, vegetables, cheese, oil, bread, cereals etc. On the other hand, higher intensity ultrasound has frequency in the range of 20 to 100 KHz with intensities more than $1W/cm^2$

and used for controlling microstructures and modification, emulsification, defoaming of food products. Ultrasound is not very effective in killing bacteria, so ultrasound is generally used with combination of other technologies like manosonication, thermosonication and manothermosonication. The ultrasonic probe system (20-40 kHz) and ultrasonic bath system (20 kHz to >1 MHz) are most commonly used systems towards the application in food industry [12].

Ultrasound treatment of dried sea cucumber (*Stichopus japonicas*) increased the ratio of rehydration and capacity to hold water with increasing levels of ultrasound power from 100 W to 300 W and decreasing levels of ultrasound frequency from 45 KHz to 28 KHz. Further ultrasound assisted rehydration increased the rehydration efficiency of dried sea cucumber by 12-fold without any adverse effects on its textural properties [13].

Lower Power Ultrasound

Lower power ultrasound (LPU) in a combination with spectroscopy and nuclear magnetic resonance are mostly used process for food analysis and the variety of wavelengths in ultrasound allows checking the properties of food and foreign bodies in food. The basic principle of LPU is that sound waves go through the food body and make compressions and decompressions. Frequency and wavelength determine the velocity of sound, in this way high-frequency sound will be shorter in wavelength and low sound have long-wavelength [1].

High Power Ultrasound

High power ultrasound (HPU) provides mechanical, biochemical, and chemical effects and it enhances the quality of many food systems in processing [14]. Mechanical shows that it has applications in the extraction of flavours, oils, degassing and destruction of foam in foods. Chemical and biochemical sterilization utilizes process-related tools to remove biofilms from the surface of food processing equipment. In HPU temperature, pressure and intensity are very important to know alongside velocity and energy.

Ultrasound Measurement Techniques

Two techniques used for ultrasound measurement are pulse-echo and continuous wave ultrasound and it is generated by transforming the electric current into ultrasound of controlled frequency. But some time pith and catch techniques were also applied to measure ultrasounds. In pulse-echo, a sample cell with a transducer and oscilloscope is used to measure ultrasound. There is a signal producer is used to make an electric pulse which converts into ultrasound pulses by the transducer. These pulses then pass through the sample and collide with the wall of the sample container, then come back to the transducer and again converts into electric signals and save on oscilloscope [15].

Possible Applications in Fish Processing

Ultrasound is currently applied in food processing like in fruits and vegetables, meat, chicken, cereals and fat products. But when we consider the use of ultrasound in fish processing, found it very limited. Many projects are under work at this point like Ultrafish, a project of EU started in Spain because of its potential in fish processing food categories [16]. Table 2 describes analysis of nutritional content of different fishes.

Table 2. Analysis of nutritional content of different fishes

Seafood	Moisture	Carbohydrates	Proteins	Fat	Ash
Bonyfish					
Bluefish	74.6	0	20.5	4.0	1.2
Cod	82.6	0	16.5	0.4	1.2
Haddock	80.7	0	18.2	0.1	1.4
Atlantic					
Halibut	75.4	0	18.6	5.2	1.0
Herring	67.2	0	18.3	12.5	2.7
Mackerel	68.1	0	18.7	16.5	1.2
Pacific					
Salmon	63.4	0	17.4		1.0
Swordfish	75.8	0	19.2	4.0	1.3

Seafood	Moisture	Carohydrates	Proteins	Fat	Ash
Crustaceans					
Crab	80.0	0.6	16.1	1.6	1.7
Lobster	79.2	0.5	16.2	1.9	2.2
Shrimp	72.5	0.9	20.5	5.5	0.8
Crayfish	80.0	0.5	17.0	1.5	0.9
Molluscs					
Clams, meat	80.3	3.4	12.8	1.4	2.1
Oysters	80.5	5.6	9.8	2.1	2.0
Scallops	80.3	3.4	14.8	0.1	1.4
Squid/mantle	83.5	1.4	13.5	0.8	0.7

Composition

The physicochemical properties of foods can be determined by using ultrasonic methods and it is also useful to determine the composition, structure and physical state of foodstuff. But these all components depend on different factors like species, age and season. Ultrasound has the advantage that it is more quick, non-destructive, accurate and automated. In ultrasonic composition analysis, it is not needed to prepare a sample as we do in general laboratory methods. Ultrasonic characteristics of fish tissues rely on temperature and composition. In solid-non-fat contents, ultrasonic velocity increases at almost all temperatures, but it becomes complex in fat areas. Commercially fillers are a very important part of fish so the water content of fish fillets can be determined rapidly and accurately by using ultrasonic velocity. The amount of fat and minerals remain constant in cod fillets which makes it easy to determine the amount of protein in cod fillets [10].

Enzyme Activity

Enzymes play a vital role in fish freshness, and it is important to control their activity to some extent so it will not harm the fish. Power ultrasound is applied to inactivate enzymes in foods and the principle of ultrasonication to inactivate the enzymes is cavitation. Cavitation provides mechanical and chemical effects against enzymes. Pressure and temperature are also used to achieve inactivation. Ultrasound produces cavitation bubbles and then by creating strong shock waves to cause strong shear force. This powerful condition allows sonication to break down the hydrogen bonding and other forces in polypeptide bonds. This will change the structure of enzymes and they will lose their activity [17]. But for different enzymes, the method of applying sonication is also different. Mostly mano-thermo-sonication (MTS), in which mild heat application with ultrasound and moderate pressure is used. MTS is very useful to those enzymes that need high temperatures for inactivation like protease. MTS requires less time and low temperature than the only thermal process for enzyme inactivation. When ultrasound is used with treatment like pressure then it is called mono-sonication and if only heat is applied then called thermo-sonication [18]. The use of ultrasonication in this field is mostly used in liquid foods section more than in solid due to cavitation process which can easily create in liquids. But mano-thermo-sonication and thermo-sonication will be of great importance in the fish industry if more research is applied to them.

Microbes And Ultrasonication

Microbes are the main cause of food-borne diseases and fish can also become hazardous to eat if it contains pathogens more than critical limits. Mostly, in industries, conventional heat process like pasteurization and sterilization is applied to get rid of microbes. These techniques successfully destroy microbes, but they also cause nutrition loss, flavour degradation, change in product properties [19]. Ultrasonication has the prospect to be applied for the inactivation of microbial populations and the mechanism of ultrasonication is that waves rapture the cell membrane, and the result is achieved. Before the application of ultrasonication, thermo-sonication, or mano-thermo-sonication. *Listeria* is a species of bacteria mainly found in fish and fish products and it is recorded that if thermo-sonication is applied on the food products having 24 kHz and 55°C for 2.5 minutes then can reduce listeria more than the conventional pasteurization process. In *E. coli*'s case, the process requires pressure also along with heating and sonication, so it then becomes mano-thermo-sonication with 20 kHz, 40-61°C, and 100-500 kPa for 0.25-4 minutes. Another important pathogen *Salmonella* can also be controlled by thermo-sonication with 24kHz and 52-58°C specifications for 2-10 minutes [20]. Table 3 lists uses of pulsed electric fields (PEF) technology on fish processing industry.

Table 3. Uses of pulsed electric fields (PEF) technology

Fish product	Operating conditions	Findings/Applications	Reference
Salmon and lumpfish roes	Different combinations of electric field, number of pulses and high-pressure treatment.	PEF application had greater impact on salmon than chicken samples.	[33]
Fishbone	Combinations of semi-bionic extraction method with PEF (optimum conditions: EFS 22.79 kV/cm and pulse number 9)	Extraction of effective ingredients	[34]
Fishbone	Processing conditions were optimized, and the best yield was achieved using a EFS of 16.88 kV/cm and pulse number of 9	Extraction of chondroitin sulfate	[35]
Pollock fillets, cod loins frozen, cod fresh fillets, haddock loins frozen, Iceland cyprine and common whelk	Electric field strength 1.2–2.0 kV/cm; frequency 1–4 Hz; pulse width 400 µs; pulse number 20, 40, 80 or 120	PEF treatment was not effective with <90 pulses and field < 2.0 kV/cm. Improvement of the water holding properties and the fish drying.	[36]
Haliotis discus hannai Inoviscera	Treatment time (100–800 µs), Intensity strength 5–20 kV/cm, and the ratio of material to solvent (3:1–10:1)	Extraction of Protein hydrolysate	[37]
Mussel	Best conditions: EFS of 20 kV/cm, pulse number of 8 and enzymolysis time of 2 h	Extraction of protein	[38]
Fishbone	Optimal combination of parameters: EFS 25 kV/cm and pulse number 8	Effective calcium extraction	[39]

Destruction of spores by thermal methods is not easy and resistant to high temperatures and pressures. Most of the time survived during processing and can cause negative effects on product shelf life. *Bacillus* and *Clostridium* spores are very resistant to pathogenic microbes. 100°C temperature for 4hours is required to destroy the spores of *Bacillus thermophiles* and manosonication is found helpful in this regard. By applying treatment at 500kPa for 12 minutes, inactivated around 99% spores of bacillus. When 20kHz is used with 300kPa and 90µm, then the inactivation of spores will be 75%. But it can increase up to 99% by just increasing amplitude to 150µm [21].

The food-borne diseases reported are found in water and most of the time do not eliminate from processing water even by treatment of that water with chlorine disinfectant technique. Ozone and ultrasound were both found very helpful to eliminate the chances of this contamination. The use of ultrasound alone in pathogens removal is not much feasible. But with a combination of ultrasound, pressure and thermal treatment more possible within a short time as compared to conventional treatments. So, the future of ultrasound as bacterial removal is relying on thermo-sonication, mano-sonication, and thermo-mano-sonication [22].

Drying

Drying is the oldest method of preservation used for fish and other food commodities and this method increases shelf life by removing water from the product. It is a vital requirement for spoilage causing microbes and enzymes to become unavailable and food remains fit for human consumption. The methods often used in fish industries are time-consuming and require days to weeks to complete procedures. Some alternative method like freeze-drying is adopted by some industries but it is very expensive and affects the price of the product. Direct or indirect contact of ultrasound applications was found suitable for food drying in a combination with mild heat [16]. The diffusion or dehydration of the product is enhanced by compression, rarefaction and pressure applied by using ultrasound waves [10].

During fish drying processing, it is noted that 43% time reduced by applying 25kHz intensity of ultrasound waves with any other technique. Just like the use of ultrasound in pathogens removal, in drying ultrasonication is applied with other techniques to achieve the target as soon as possible. So, the influence of ultrasound depends on the selection of different parameters for drying like mode of ultrasound application, other combined drying techniques, the intensity of waves, conditions applied, and type of material [23]. Fish drying will become safer to do if ultrasound is applied during drying as it will cut down the processing time and the chance of quality deterioration of fish products will be minimum with no side effects at low cost.

Freezing

Freezing is also a very popular method of preservation in the food sector and many fish processes demand frozen fish to be delivered as raw material.

Freezing has a very good effect on product quality and shelf life. The loss of nutrients and flavour is much lower than other techniques. Freezing is done by converting water into crystals and quick-freezing method helps to gain small crystals, but the slow methods create large crystals that will damage the structure of the product. But the rapid freezing techniques, very expensive to apply and it also increases the price of products. High power ultrasound is used successfully to control crystallization and nucleation procedure, it also remains noncontact and has no side effects on the product chemically [24]. HPU is still not applied in fish industries, but it is used in freezing with slow freezing methods. Thawing is a very critical and slow step in frozen fish processing. HPU application on cod thawing gives very good results. Cod flesh received 500kHz frequency with 0.5W intensity which takes 2.5 hours for thawing. This time is less than the usual water methods and it also did not change the quality characteristics of cod. In another case, 1500kHz, HPU was applied on the cod block with 60W intensity, and it helps to decrease 71% time as compared to usual water methods.

Brining

Brining is a process in which salt of the solution goes into the flesh and water of the flesh comes out. It is a very vital step in fish processing for many ready-to-eat products. It does not only add taste but also gives a preservation effect. Simple bringing solution applications take much time, and we need to use more salt. Conventional brining is not much remained under control as it is not performed uniformly on the whole product. It is noted in flesh that saturated NaCl solution with different ultrasonication energy with great agitation, showed that the NaCl penetration is higher in sonicated than non-sonicated. It also reduces the penetration time of NaCl. The brining assisted with ultrasound also helps to distribute brine in all the products equally, but this all depends on the size of flesh, sound energy, and agitation frequency. The mass transferability of ultrasound also shows that it can be useful in fish brining and other marination procedures which require a lot of time to get ready for final processing and packaging [25].

Extraction by Ultrasound

Typically, fish oil and flavour are extracted by harmful chemicals and high energy costs. This method has many drawbacks and the concern of extraction technology converted towards green extraction technologies assisted by harmless technologies like microwave, ultrasound, ultrafiltration, and flash distillation process. Ultrasound is becoming very popular in extraction methods, and it is called the ultrasound-assisted extraction (UAE) method used with high power ultrasound. The mechanism of UAE is that when waves are applied on the medium it creates compression and rarefaction.

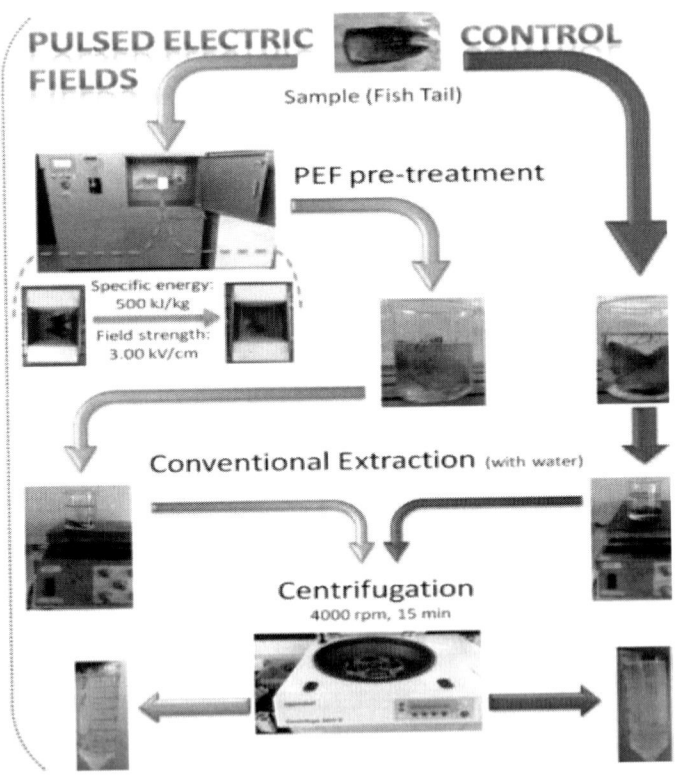

Figure 1. The extraction process of hydrophilic compounds from a fish by-product (fish tail) with and without pulsed electric field (PEF) pre-treatment.

So, by this application, oil is extracted from fish without the use of any chemical material. Asian swamp eel is used in an experiment to extract oil with the help of the UAE method. The parameters set for the procedure are

25kHz, 200W, and 60 minutes with only 500 ml of ethanol. The procedure was recorded successfully. In another work on the same species was done with the same frequency but 400W, 50 °C, and for 57 minutes. It gave 94.82% extraction [20]. Figure 1 shows the extraction process of hydrophilic compounds from a fish by-product (fish tail) with and without pulsed electric field (PEF) pre-treatment.

Other Applications

Ultrasound waves can be used infiltration to avoid blockage of the filter membrane. It makes solid particles move around and this way they will not settle down and it will not create on the membrane. This filtration assistance can be applied in fish sauce processing and other liquid filtrations during processing. Fish fillet cutting is a process of high concern regarding product quality. If cutting of fillets is not accurate as wanted for the product it could become a drawback during processing. Conventional knives produce more waste and are less accurate. Although in fish processing, high-pressure water jet cutting becoming more popular, the energy consumption of the ultrasound method is far less than water jet cutting [26].

Challenges with Ultrasound Application

The main challenge in the application of ultrasound is the lack of standardization for the different procedures in fish and food processing at the commercial-scale level and only 20-40kHz frequency is available still in use for the commercial level. In the liquid phase when applied small gas bubbles make it difficult for the waves to propagate through the medium [27]. It is also difficult to convince consumers that ultrasound-treated foods are not harmful for consumption.

Regulations on the Use of Emerging Technologies in Bio-Marine Food Products

With almost 7.8 billion people in a complicated landscape of climate change, an increased competition for natural resources, and economic and financial uncertainty, it is necessary to establish goals for the contribution and conduct of fisheries, aquaculture, and the whole bio-marine sector toward food safety and nutrition in order to ensure sustainable economic, social, and environmental development. To do so, it is also important to create a legislative framework in order to control and supervise all new advances in the sector ensuring the consumer's protection in each step of the food chain. Therefore, current regulation on the use of novel bio-marine resources and new processing technologies from an international perspective are urgently needed [11].

Need for Innovative and Emerging Technologies in the Bio-Marine Food Sector

Regulations and prospects of i*nnovative and emerging technologies* presents the use of technologies and recent advances in the emerging marine food industry. It includes technological design, equipment and applications of these technologies in multiple processes. Extraction, preservation, microbiology and processing of food are extensively covered in the wide context of marine food products, including fish, crustaceans, seafood processing waste, seaweed, microalgae and other derived by-products. It is an interdisciplinary resource that highlights the potential of technology for multiple purposes in the marine food industry as these technological approaches represent a future *alternative to develop more efficient industrial processes* [28].

Pulsed Electric Fields for the Extraction of Proteins and Carbohydrates from Marine Resources

Pulsed electric fields (PEF) are attracting great attention as non-thermal extraction technology on downstream processing of emergent marine bioresources such as micro- and macroalgae. These uni- and multicellular algae enclose a diverse and interesting biochemical composition and are recognized as important sources of valuable bioactive compounds and functional food nutrients. The role of electric field processing towards the extraction of proteins and carbohydrates given particular emphasis on PEF technology, addressing the latest developments in the electric field-based technologies. It is noted as a great potential as a "green" tool, being an energy efficient and environmentally friendly processing technology, by reducing the operational time by allowing a strong electro permeation, that is electroporation of cellular membranes, that enhances the extraction yields while reducing the use of solvents. The possibility of combining PEF with other processing variables can bring more flexibility to the extraction process [29].

Supercritical Fluid Extraction of Lipids, Carotenoids, and Other Compounds from Marine Sources

Research on high value compounds such as lipids, including polyunsaturated fatty acids, sterols, phospholipids, as well as carotenoids and phenolic compounds has raised great attention, due to the increasing interest of consumers in functional foods which can promote health benefits. Microalgae, macroalgae, fisheries products and by-products have shown to be a great potential source of those recognized bioactive compounds. In this sense, interesting developments have been carried out employing supercritical fluid extraction (SFE). In addition, advances on gas-expanded liquid and new bio refinery approaches using SFE for obtaining bioactive compounds enriched extracts are presented. Finally, a perspective about the potential applications of the marine bioactive extracts in the food industry is included [20].

Microwave-Assisted Extraction of Proteins and Carbohydrates from Marine Resources

Microwave-assisted extraction (MAE) is an efficient method that uses microwave energy to accelerate the removal of bioactive compounds from natural matrices. Marine resources are a potential source of bioactive compounds, which has wider applications in food, pharmaceutical, and cosmetic industry. Under the specified conditions, electromagnetic microwaves efficiently penetrate the material structure, generating a distributed heat source that promotes the extraction of bioactive compounds from the marine biomass. This chapter reviews the application of the MAE technique to extract polysaccharides and proteins from marine resources such as microalgae, macroalgae, and fish [30].

Microwave-Assisted Extraction of Lipids, Carotenoids, and Other Compounds from Marine Resources

Marine resources are rich in biologically active substances including lipids, carotenoids, polyphenols, etc., and are widely used in food, medical, and other industries [1]. Microwave-assisted extraction (MAE), as an efficient, green advanced extraction technology, exhibits better heating efficiency and less energy consumption than traditional extraction methods. As a promising wet lipid extraction process, the extraction efficiency of MAE is mainly depended on the operating parameters and the types of raw materials. The extraction of carotenoids by the MAE method is mainly focused on the extraction of astaxanthin, fucoxanthin, and β-carotene. In addition, MAE has been widely studied for the extraction of other biologically active substances such as polyphenols and phycobiliproteins. The joint use of MAE and other extraction technologies is also widely applied and has great potential and applications and trends of MAE to extract important compounds such as lipids, carotenoids, polyphenol, phycobiliprotein from marine resources.

Extraction of High-Value Compounds from Marine Biomass via Ionic Liquid-Based Techniques

Ionic liquids (ILs) have been suggested as promising media to separate and extract bioactive compounds from a broad range of natural feedstocks. The unique physicochemical and solubilization properties of ILs have also led to new developments in separation science and materials science in recent years. This review highlights recent accomplishments in extraction processes of diverse high-value compounds from different kinds of marine biomass such as fish and marine algae via the use of ILs. High-value products targeted in recent studies through ILs-based processes include lipids, small organic extractable compounds, proteins, etc. Industrial applications of ILs for extraction processes often employ combinations with traditional organic solvents.

Application of Pressurized Liquids to Extract High-Value Marine Compounds

Pressurized liquid extraction (PLE) is a promising green technology to extract various value- added compounds from marine biomass. In principle, the PLE technique employs pressurized solvents (typically 5–20 MPa) at high temperatures (generally 80–200°C), which allows an exhaustive extraction of intracellular compounds in short time. The technique dramatically reduces solvent consumption compared to conventional extraction processes. The solvent choice varies depending on the chemical nature of the target compounds, although water and ethanol have shown to be the most suitable solvent complying green chemistry principles. The optimal temperature varies depending on the target compound, while 5–20 min is sufficient to extract most compounds. Thus, it provides the principles and application of PLE technology for the extraction of high-value compounds from marine biomass [31].

Application of Plasma Technologies for Food Preservation

Consumers' demand for high-quality and microbiologically safe seafood products has added value to emerging nonthermal plasma technologies and is attracting the attention of researchers in the food industry. Cold plasma (CP) has shown the ability to inactivate a wide range of common foodborne pathogens on seafood products and thus can help maintain quality attributes and nutrient value. Thus, it enhances the principles governing the functionality of plasma and recent developments of cold plasma technologies application on marine products, including preservation purpose and marine proteins alteration. The microbiological safety and quality of multiple marine-derived products will be analysed with this plasma technology in the food processing industry [32].

Conclusion

The various health benefits of seafood species over red meat is one of the main reason behind the increasing consumption of seafoods and it has become one of the major source of animal protein in the human diet. During past decades the aquaculture has grown at a very high rate and the fish or seafood is becoming more accessible to the consumers. So, by keeping this factor in mind and the perishable nature of seafoods, there is a greater need for various innovative and safer technologies for the processing of seafoods. The various innovative techniques employed for seafood processing have a tremendous potential for extending the shelf life, reducing the wastage, and delivering quality food in terms of nutritive properties. Ultrasound has attracted considerable interest in food science and technology due to its promising effects in food processing and preservation. As one of the advanced food technologies, it can be applied to develop gentle but targeted processes to improve the quality and safety of processed foods. It also offers the potential for improving existing processes as well as for developing new process options. There are an increasing number of industrial processes that employ power ultrasound as a processing aid and can be a specialized as a versatile technology with numerous applications in food processing. Ultrasonic processing is still in its infancy stage in marine sector and requires a great deal of future research in order to develop the technology on an industrial scale, and to more fully elucidate the effect of ultrasound on the properties of foods. Still, much potential research is needed to develop industrial-automated ultrasound systems to help minimize labour, prices and resources. Ultrasound is an innovative technology with efficiency, non-toxic and eco-friendly and ultrasound waves provide non-destructive, easy application and fast processing procedures. This technique will not harm the environment and can use it as a preservation technique by applying it against microbes and enzymes. The approach of ultrasound against microbes is destructive and its inactivation ability of enzymes without any quality and nutritional deterioration makes it more acceptable for new generation value-added products. Although this method has not completely become an alternative to conventional methods still this can make procedures much easier and reduce time by applying it with a combination of different conventional techniques. Ultrasound has more benefits but still, a lot of studies are needed to minimize its capital investment, its scale-up standardization and other challenges associated with it.

References

[1] Awad, T. S, Moharram, H. A, Shaltout, O. E, Asker, D, Youssef, M. M. Applications of ultrasound in analysis, processing and quality control of food: A review. *Food Res. Int,* 2012, 48: 410-427.

[2] Balasubramanian, S, Balasubramaniam, V. M. Compression heating influence of pressure transmitting fluids on bacteria inactivation during high pressure processing. *Food Res. Int*, 2003, 36: 661–668.

[3] Arvanitoyannis, I. S, Kotsanopoulos, K. V, Savva, A. G. Use of Ultrasounds in the Food Industry- Methods and Effects on Quality, Safety and Organoleptic Characteristics of Foods: A Review. *Food Science and Nutrition*, 2015, 57(1), 109-128.

[4] Belitz, H. D, Grosch, W, Schieberle, P. Cereals and cereal products. *Food chem*, 2009, 670-745.

[5] Boskou, G, Debevere, J. In vitro study of TMAO reduction by Shewanella putrefaciens isolated from cod fillets packed in modified atmosphere. *Food Additives and Contaminants*, 1998, 15: 229-236.

[6] Carver, C. H. Vacuum cooling and thawing fishery products. *Mar. Fish. Rev*, 1975, 37: 15-21.

[7] Chandrasekaran, S, Ramanathan, S, Basak, T. Microwave food processing—A review. *Food Res. Int*, 2013, 52: 243-261.

[8] Cheftel, J. C, Culioli, J. Effects of high pressure on meat: a review. *Meat sci*, 1997, 46: 211-236.

[9] Day, B. P. Active packaging of food. Smart packaging technologies for fast moving consumer goods. 1-18. De la Fuente-Blanco S., De Sarabia E. R. F., Acosta-Aparicio V. M., Blanco-Blanco A. and Gallego-Juárez J. A. 2006. Food drying process by power ultrasound. *Ultrasonics*, 2008, 44: 523-527.

[10] Boateng, E. F, Nasiru, M. M. Applications of Ultrasound in Meat Processing Technology: A Review. *Food Science and Technology*, 2019, 7(2), 11-15.

[11] Ravishankar, C. N. Advances in Processing and Packaging of Fish and Fishery Products. *Advanced Agricultural Research & Technology Journal*, 2019, 3(2), 168-181.

[12] Creed, P. G. Sensory and nutritional aspects of sous vide processed foods. *Sous vide and cook-chill processing for the food industry*, 1998, 57-88.

[13] Church, I. J, Parsons, A. L. The sensory quality of chicken and potato products prepared using cook–chill and sous vide methods. *Int. J. Food Sci. Tech*, 2000, 35: 155-162.

[14] Jayasooriya, S. D, Bhandari, B. R, Torley, P, D'Arcy B. R. Effect of High-Power Ultrasound Waves on Properties of Meat: A Review. *International Journal of Food Properties*, 2004, 7(2), 301–319.

[15] Buckin, W, Kudryushov, E, O'Driscoll, B. High-resolution ultrasonic spectroscopy for material analysis. *American Laboratory*, 2002, 28–31.

[16] Charoux, C. M. G, Ojha, K. S, O'Donnell, C. P, Cardoni, A, Tiwari, B. K. Applications of airborne ultrasonic technology in the food industry. *Journal of Food Engineering*, 2017, 208: 28-36.

[17] Mawson, R, Gamage, M, Terefe, N. S, Knoerzer, K. Ultrasound in Enzyme Activation and Inactivation. *Ultrasound Technologies for Food and Bioprocessing*, 2010, 369–404.

[18] Vercet, A, Burgos, J, Crelier, S, Lopez-Buesa, P. Inactivation of proteases and lipases by ultrasound. *Innovative Food Science & Emerging Technologies*, 2001, 2(2), 39–150.

[19] Dolatowski, Z. J, Stadnik, J, Stasiak, D. Applications of ultrasound in food technology. *Acta Science*. 2007, 6(3), 89-99.

[20] Ivanovs, k, Blumberga, D. Extraction of fish oil using green extraction methods: a short review. *Energy Procedia*, 2017, 128: 477–483.

[21] Ghaedian, R, Coupland, J. N, Decker, E. A, McClements, D. J. Ultrasonic determination of fish composition. *Journal of Food Engineering*, 1998, 35(3), 323–337.

[22] Piyasena, P, Mohareb, E, McKellar, R. C. Inactivation of microbes using ultrasound: a review. *International Journal of Food Microbiology*, 2003, 87: 207–216.

[23] Musielak, G, Mierzwa, D, Kroehnke, J. Food drying enhancement by ultrasound – A review. *Trends in Food Science & Technology*, 2016, 56:126–141.

[24] Miles, C, Morley, M, Rendell, M. High power ultrasonic thawing of frozen foods. *Journal of Food Engineering*, 1999, 39(2), 151–159.

[25] Gallego-Juárez, J. A, Rodriguez, G, Acosta, V, Riera, E. Power ultrasonic transducers with extensive radiators for industrial processing. *Ultrasonics Sonochemistry*, 2010, 17: 953- 964.

[26] Faustman, C, Cassens, R. G. The biochemical basis for discoloration in fresh meat: a review. *J. Muscle Foods*, 1990, 1: 217-243.

[27] Rastogi, N. K. Opportunities and Challenges in Application of Ultrasound in Food Processing. *Critical Reviews in Food Science and Nutrition*, 2011, 51:705–722.

[28] Farber, J. M. Microbiological aspects of modified-atmosphere packaging technology-a review. *J. Food Protection*, 1991, 54: 58-70.

[29] Dossat, R. J. Principles of refrigeration (3rd ed). Englewood Cliffs, NJ, USA: Prentice Hall. Doyle M. E. 1999. Food irradiation. Madison: University of Wisconsin, Food Research Institute. Dunn J., Ott T. and Clark W. 1995. Pulsed light treatment of food and packaging. *Food Tech*, 1991, 49: 95–98.

[30] Chemat, F, Zill-e-Huma, Khan, M. K. Applications of ultrasound in food technology: Processing, preservation and extraction. *Ultrasonics Sonochemistry*, 2011, 18(4), 813–835.

[31] Datta, A. K, Anantheswaran, R. C. Handbook of Microwave Technology, for Food Applications., Eds. Marcel Dekker, New York. Datta A. K. and Davidson P. M. 2000. Microwave and radio frequency processing. *J. Food Sci*, 2001, 65: 32-41.

[32] Daniels, J. A, Krishnamurthi, R, Rizvi S. S. A review of effects of carbon dioxide on microbial growth and food quality. *J. Food Protection*, 1985, 48: 532-537.

[33] Gudmundsson, M, Hafsteinsson, H. Effect of electric field pulses on microstructure of muscle foods and roes. *Trends in Food Science & Technology*, 2001, 12(3–4).

[34] He, G, Yin, Y, Yan, X, Wang, Y. Semi-bionic extraction of effective ingredient from fishbone by high intensity pulsed electric fields. *Journal of Food Process Engineering*, 2017, 40(2), e12392.

[35] He, G, Yin, Y, Yan, X, Yu, Q. Optimisation extraction of chondroitin sulfate from fish bone by high intensity pulsed electric fields. *Food Chemistry*, 2014, 164, 205–210.

[36] Kumar, Y, Kumar Patel, K, Kumar, V. Pulsed electric field processing in food technology. *International Journal of Engineering Studies and Technical Approach*, 2015, 1(2), 6–16.

[37] Li, M, Lin, J, Chen, J, Fang, T. Pulsed electric field-assisted enzymatic extraction of protein from abalone (Haliotis Discus Hannai Ino) viscera. *Journal of Food Process Engineering*, 2016, 39(6), 702–710.

[38] Zhou, Y, He, Q, Zhou, D. Optimization extraction of protein from mussel by high-intensity pulsed electric fields. *Journal of Food Processing and Preservation*, 2017, 41(3), e12962.

[39] Zhou, Y, Sui, S, Huang, H, He, G, Wang, S, Yin, Y, Ma, Z. Process optimization for extraction of fishbone calcium assisted by high intensity pulsed electric fields. *Nongye Gongcheng Xuebao/Transactions of the Chinese Society of Agricultural Engineering*, 2012, 28(23), 265–270.

Chapter 6

Comparison of Antioxidants and Ultrasound Treatment on the Color, Enzymatic Browning and Bioactive Compounds of Fresh-Cut Radishes

Arzu Imece*
Department of Food Engineering, Igdir University,
Iğdır, Turkey

Abstract

The impact of ultrasound (US) and antibrowning agents (ascorbic acid & citric acid) on the minimization of enzymatic browning, color and bioactive compounds for 10 days shelf-life study of fresh-cut radish samples was determined. The treated fresh-cut radish pieces were analyzed in terms of color, polyphenol oxidase, pectin methyl esterase activity and selected bioactive components (ascorbic acid, antioxidant activity and total phenolics).

The US pre-treated radish pieces exhibited a significant ($p < 0.05$) development in bioactive components. In a similar way, US pre-treated radish slices displayed the best color parameters with the lowest value related to enzyme activity meaning less enzymatic browning. It is declared that US pre-treatment is an alternative technique for the replacement of chemical applications to extend the storage conditions and keep the quality characteristics in fresh-cut radishes during cold storage.

---

* Corresponding Author's Email: arzu.odunkiran@igdir.edu.tr.

In: Power Ultrasound and Its Applications in Food Processing
Editors: Gulcin Yildiz and Gökçen Yıldız
ISBN: 979-8-88697-639-7
© 2023 Nova Science Publishers, Inc.

Keywords: antioxidants, bioactive compounds, radish, shelf-life, ultrasound treatment

Introduction

In the fresh cut fruit and vegetables sector, enzymatic browning has a really crucial role. This kind of browning degradation causes reduction in the shelf life of various processed food products, but also inversely influences the quality characteristics of fruit and vegetable crops. Because of enzymatic browning occurrence approximately fifty percent of tropical fruits are eliminated due to the quality faults [1]. The browning process is essentially catalyzed by the specific enzyme named as PPO (polyphenol oxidase) which is also called as phenol oxidase, phenolase, monophenol oxidase, diphenol oxidase and tyrosinase [2]. When fruit and vegetable tissues are injured because of the physical factors including cutting/slicing, bruising or blending, PPO becomes an active form as a result of releasing into the cytosol. With the existence of molecular oxygen and PPO, monophenol is hydroxylated into o-diphenol and diphenol might be oxidized into o-quinones, which then goes through polymerization phenomena in order to create dark brown polymeric substances.

For the inhibition of this phenomenon, several methods are developed consisting of thermal, chemicals (particularly ascorbic acid (AA)), and physiological techniques. Antioxidants or reducing agents inhibit the browning process with the help of reducing oquinones which is formed enzymatically into the corresponding diphenols with no colour. They could respond irreversibly with the oquinones, producing more durable components with no color, including ascorbic acids and its derivatives. The useful impacts of ascorbic acids are associated with various attitudes i.e., creating a chemical barrier which inhibits the O_2 diffusion into the food material, hence, lessen the generation of oquinones and preventing the PPO [3]. In addition, citric acid (CA) can also be utilized as an antioxidant agent, showing synergistic relationship with the ascorbic acid which might cause rancid taste and/or flavor and enzyme inactivation including the PPO [4]. The ascorbic acid content of almost all vegetable crops reduces while bruising, trimming, cutting and slicing happens [5].

Moreover, although thermal processes are very efficient in order to minimize enzymatic browning process, it is the main reason of adverse changes occurred in the appearance, structure, color, and flavor of fruit and vegetable crops [6]. Therefore, non-thermal technology is searched as an alternative in order to retain the fresh-like color, and texture of fresh-cut food materials [7] in addition to effective on enzyme inactivation such as high pressure decreases the PPO activity (37%) in ginger [8], complete elimination of browning enzymes in apple subsequent to pulsed electric field treatment [9], total inhibition of other enzymes following thermo-sonication process in grapefruit juices [10].

The system behind the enzyme inactivation might be the appropriate usage of temperature degree, ultrasound frequency, treatment duration, and the food structure [11].

Even though being a vegetable of less importance depending on the cultivated area, radish (Raphanu sativus L.) is believed important for several small property owners who grow a great variety of vegetables. Moreover, it is an outstanding resource of calcium, phosphor and mangan, includes vitamins B_1 and B_2, nicotin acid and ascorbic acid, acts as a diuretic and antiscorbutic preservatives, and provokes the digestive granules and liver – advancing better digestion [12].

On the other hand, radish is acquiring a good market share of minimally processed food products, even though its physical behavior after packaging is still not known very well. In this study, the purpose was to compare 2 well-known antioxidant agents, namely ascorbic acid and citric acid, their combination and ultrasound on the inhibition of enzymatic browning in radish as a result of minimal processing.

Materials and Methods

Preparation of Samples

Fresh radish samples were purchased from Iğdır local market, Turkey. Radishes are cut into 3-cm-long quarters along by a sharp knife and exposed to 6 treatments in total right after slicing radish pieces. The sample with no pretreatment was expressed as a control. For the antioxidants, radish slices were submerged into antibrowning agents (ascorbic and citric acids) for about 300 seconds.

All the antibrowning agents (AA & CA) were arranged at 3% (w/v) with distilled water in addition to AA and CC combination at 1.5% AA-1.5% CC. Radishes were immersed in an ultrasonic water bath (Wiseclean, WUC-A10H, German) for 15 minutes while applying ultrasound pre-treatment. Following the application of all methods, fresh-cut radish slices were drained for one minute with a paper towel and immediately stored in polyethene bags, labelled, and held at refrigeration temperatures for 10 days for shelf life analysis at 0, 5 and 10 days.

Color Measurement

A Konica Minota (CR-40.0, Osaka, Japan) with illuminant D65 and 8 mm measurement space in the CIE $L^* a^* b^*$ color scope was organized to observe the color parameters of fresh-cut radishes. The colormeter was then calibrated by a standard white and black surface, subsequently. Lightness (L^*), redness (a^*, red-green), and yellowness (b^*, yellow-blue) were used to characterize the color values. L^*, a^*, and b^* parameters were obtained at 10 separate places at the external part of the surface at 25 °C and the averaged numbers (L^*, a^*, and b^*) were described.

PPO Activity and PME Activity

The technique already defined by Yildiz et al. [12] was conducted to measure PPO activity of fresh-cut radish slices with a spectrophotmeter (Cary 60 UV-VIS, Agent Technologie, USA), while PME activities of fresh-cut radish pieces were determined as reported by the procedure of Cruz-Casino et al. [13].

Ascorbic Acid Content (AAC, Vitamin C)

AAC of radish samples was obtained by titration method according to the procedure reported by Rangana [14] depending on the decrease of 2,6-dichloroindophenol dye (SigmAldrich, St. Louis, MO, USA) by ascorbic acid, and it was shown as mg/100 g of fresh fruit mass.

Sample Extraction

The extraction phase was managed according to the technique mentioned by Yildiz et al. [12] Fresh and treated radish samples (1 g) were homogenised and mixed with 4.5 mL CH3OH and water (80/20 v/v) at 25 °C and shaken for 120 minutes at 140 rpm using an orbital shaker (Biosan OS-20, Lativa). The supernatant was collected after centrifugation (10000g, 15 min) (Sigma 3K-30, UK).

The supernatant was filtered (0.45 μm) PTFE filter to determine the total phenolic compounds and ATC of radish pieces.

Total Phenolic Content (TPC) and Antioxidant Capacity (ATC)

The analysis explained by Igual et al. [15] along with slight modification was conducted for the measurement of TPC of radish slices by gallic acid as a standard. While DPPH analysis was applied for the determination of the ATC on radish slices based on the approach explained earlier by Yildiz & Aadil [16].

Statistical Analysis

A randomized plot factor experimental design was managed. The outcomes were obtained via the JMP (Version 7.0, SAS Institute Inc., Cary, NC, USA). Differences between averaged numbers were determined through Fisher's LSD test (0.05).

Results and Discussion

Color Measurement

The color values of both fresh and processed radish samples (chemicals and ultrasound) were tabulated in Table 1. US-treated radish samples showed a significantly higher $L*$ value (all days) as compared to chemical treatment after 10 days of storage (Table 1). The physical degradation and membrane destruction of the cells led by ultrasound application caused simple elution of color components from fruit composition [17]. As the storage period was extended (day 0 to 10), the $L*$ parameter was significantly decreased ($p < 0.05$) in all radish pieces (untreated & processed radish slices). The $a*$ values of the radishes exhibited another way progression with the $L*$ parameters and increased during the storage period (Table 1). Whilst the lowest $a*$ parameter was observed on the first measurement (day 0) for all radish pieces, the highest a^* parameter was determined for the radish samples on day 10. In addition, b^* parameter of the radish samples was increased as the storage period in all treated radishes. While the highest b^* parameter was determined for the 10th days radish samples, the lowest b^* parameter was observed for the radish slices at day 0. Color is a significant factor that affects people choices and lets consumers to get an idea for several properties related to final food products [18]. It is stated that an increase in b^* value is sign of browning activity [19]. In our study, the b^* parameters were increased during 10-days of cold storage showing the browning development. Kasım and Kasım [20] observed similar results where the colour indexes of fresh-cut green bean samples demonstrated an increase in b^* parameters during the shelf-life study (11 days). In addition, color values of fresh-cut Chinese water chestnut samples [18] and pineapple samples [21] were recorded by an increase in b^* parameter with shelf life period (12 days). US-treated radish samples showed the lowest increase in b^* values, which shows the less browning occurrence (Table 1). In shortly, the US application is more promising to retain the color of radish, which shows good-quality and extended product life of food materials.

Table 1. Change of color values in control and treated fresh-cut radish samples over storage at 4°C

Treatment	Storage (Day)	L^*	a^*	b^*
Control	0	75.25 ± 0.15 [b (z)]	-0.82 ± 0.01 [a (x)]	3.19 ± 1.71 [a (z)]
	5	53.76 ± 0.22 [b (y)]	3.53 ± 0.53 [b (y)]	6.51 ± 0.84 [a (y)]
	10	48.18 ± 1.08 [d (x)]	3.62 ± 1.81 [c (y)]	7.34 ± 0.13 [a (x)]
Ascorbic acid (3%)	0	75.92 ± 1.19 [b (z)]	-0.83 ± 0.07 [a (x)]	3.12 ± 1.76 [b (y)]
	5	52.95 ± 0.42 [b (y)]	4.27 ± 0.19 [b (y)]	5.75 ± 0.24 [c (y)]
	10	51.13 ± 1.76 [b (x)]	4.18 ± 0.24 [c (y)]	5.35 ± 1.22 [c (x)]
Citric acid (3%)	0	75.38 ± 0.07 [c (z)]	-0.87 ± 0.06 [a (x)]	3.65 ± 1.27 [b (z)]
	5	54.25 ± 1.28 [a (y)]	6.46 ± 0.47 [b (y)]	4.99 ± 0.38 [c (y)]
	10	52.65 ± 0.55 [b (x)]	6.45 ± 1.74 [b (z)]	4.85 ± 0.19 [c (x)]
AA (1.5%)-CA (1.5%)	0	75.63 ± 0.15 [b (z)]	-0.83 ± 0.09 [a (x)]	3.66 ± 1.08 [c (y)]
	5	55.14 ± 0.33 [a (y)]	8.24 ± 1.23 [a (y)]	4.83 ± 0.54 [d (y)]
	10	53.28 ± 1.46 [b (x)]	10.15 ± 0.38 [b (y)]	4.79 ± 0.28 [d (x)]
Ultrasound (15 min)	0	75.12 ± 0.29 [a (z)]	-0.88 ± 0.02 [a (x)]	3.01 ± 1.63 [d (z)]
	5	65.47 ± 1.31 [a (y)]	15.13 ± 1.38 [a (y)]	3.25 ± 0.16 [e (y)]
	10	56.39 ± 1.16 [a (x)]	19.21 ± 0.31 [a (z)]	3.27 ± 0.43 [e (x)]

[a-f] Treatment means showed the effect of different treatments for the same day are not significantly different (p<0.05).

[x-z] Treatment means showed the effect of storage times for the same treatment are not significantly different (p<0.05).

Enzymatic Activity (PPO and PME)

The PPO and PME activities of fresh-cut radish pieces during 10 days at 4°C are demonstrated in Table 2. The PPO activities of radish samples by US displayed the lowest values during the 10 days of the shelf-life period (Table 2). The PPO activity of all radish pieces was significantly increased ($p<0.05$) after the storage interval, especially for fresh radish slices. While PPO activities of radish slices were low on day 0, they were shown higher values on other days (5 and 10). So, the highest PPO activity were determined for the stored radish slices at day 10 (Table 2). More importantly, radish slices exposed to ultrasound treatment displayed the lowest PPO activity with the comparison of the control and other treated radish slices indicating less browning. On the other hand, it was figured out that the PME activities of all treatments including fresh radish slices were significantly increased after 10 days of storage period ($p<0.05$). While PME activity of radish slices was found lower for the first-day samples, it was higher for the samples on the other days (5 and 10).

Table 2. Changes of PPO activity in fresh-cut radish slices at 4°C

Treatments	Storage (Days)	PPO activity (U/mL)	PME (UPE/ml)
Control	0	635.66 ± 0.38 a(z)	15.45 ± 1.28 a(z)
	5	763.23 ± 1.76 a(y)	16.13 ± 0.19 a(y)
	10	911.15 ± 2.09 a(x)	18.28 ± 1.07 a(x)
Ascorbic acid (3%)	0	575.87 ± 0.18 b(z)	14.25 ± 1.08 c(z)
	5	632.12 ± 2.43 c(y)	15.33 ± 1.14 b(y)
	10	723.99 ± 1.66 d(x)	16.47 ± 0.12 c(x)
Citric acid (3%)	0	538.78 ± 0.42 c(z)	14.44 ± 0.28 b(z)
	5	697.94 ± 0.74 b(y)	15.41 ± 0.81 b(y)
	10	801.15 ± 0.15 b(x)	16.92 ± 0.14 c(x)
AA (1.5%) - CA (1.5%)	0	535.26 ± 1.35 c(z)	13.76 ± 1.52 d(y)
	5	692.57 ± 0.97 b(y)	14.67 ± 1.61 c(x)
	10	764.21 ± 2.05 c(x)	15.37 ± 1.78 d(x)
Ultrasound (15 min)	0	337.13 ± 1.07 d(z)	11.13 ± 0.69 e(y)
	5	425.41 ± 0.78 d(y)	12.25 ± 0.34 d(x)
	10	454.66 ± 0.15 e(x)	13.12 ± 0.09 e(x)

[a-e] Treatment means showed the effects of different treatment for the same day are not significantly different ($p<0.05$).

[x-z] Treatment means showed the effect of storage times for the same treatment are not significantly different ($p<0.05$).

So, the highest PME activity were found at the last-day radish samples (Table 2). Overall, as compared to all other samples, US-treated radish slices had the lowest PPO and PME activity (p<0.05), indicating reduced browning, which is supported by the color values (Table 1). Fresh-cut radish slices subjected to ultrasound treatment demonstrated less browning with the lowest $b*$ values during the storage period as compared to control and other treated radish slices.

It was discovered that the employment of ultrasound application restricts the activity of PPO and peroxidase in potato [22] and cucumber [23]. The skin removal of fruits was caused by food processing procedures such as cutting, and this is also the case for weight loss and degradation. It can also lead to higher respiration rates and ethylene formation [24].

In addition, the interface between PPO and phenol directed to the browning [25]. The cavitation created after ultrasound waves, which is crucial for the formation of microscopic channels in the fruit and/or vegetable samples that assists the progress of moisture removal [26, 27]. Furthermore, this could be useful for removing moisture that has become firmly bonded to solid materials exposed to ultrasonic waves.

Because of the presence of microscopic channels, porous deformation of the solid food products was generated after ultrasound treatment reduces the diffusion boundary layers and promotes the convective mass transfer in food materials [28].

Bioactive Compounds

The ascorbic acid content (AAC) of radish samples for 10-days of storage is shown in Table 3. A significant decline for AAC was found in control and all treated radish slices during storage (p < 0.05). After 10 days of storage time, AAC of radish slices subjected to US was significantly higher (p<0.05) with the comparison of control radish slices. A decrease in AAC of mango [29] and cucumber slices [23] during 15-days of storage were also reported. The decrease in AAC as a function of time was observed frequently in processed fruits and vegetables. Their involvement in the oxidative reaction is probably the main reason for this [30]. Even though the decrease in AAC of radish slices exposed to US application was determined, the loss was found lower as compared to other samples (Table 3). The removal of oxygen caused by the cavitation effect might be the reason for ascorbic acid loss [31].

The effects of TPC and ATC of control and processed radish samples are shown in Table 3. US-treated fresh-cut radish slices demonstrated the highest phenolic compounds starting from the first day to the last day (Table 3). This might be caused by the release of the bound form of phenolic components as a result of the cell wall deformation thanks to ultrasonic cavitation [32]. Comparable observations were reported in the researches of Bhat et al. [33] and Cruz-Cansino et al. [13]. The ATC of radish samples was significantly increased (p<0.05) after storage period for all treatments (Table 3). Radish pieces subjected to ultrasound showed the highest ATC between the treatments. Similarly, enhanced bioactive compounds and ATC of tomatoes during storage were observed in the work of Lu et al. [34]. The improved extraction efficiency of antioxidant chemicals such as vitamin C and phenolic substances, which is related to the mechanical impact of cavitation during power ultrasound application, could explain the increase in ATC [35]. This supported by the findings where there is a positive interaction between ATC and TPC of radish slices (Table 3). While the lowest ATC and TPC were found for the control samples in the whole storage period, the highest ATC and TPC were determined for the fresh-cut radish slices exposed to power ultrasound treatment.

Table 3. The effect of AAC, ATC, and TPC in fresh-cut radish slices at 4°C

Treatment	Storage (Day)	AAC (mg/100 g)	ATC (μmol TE/g)	TPC (mg GA/E100g)
Control	0	20.08±0.07$^{d(x)}$	6.64±0.13$^{a(z)}$	308.15±0.46$^{c(y)}$
	5	14.17±0.89$^{c(y)}$	7.62±0.28$^{c(y)}$	310.75±0.23$^{d(x)}$
	10	8.33±0.14$^{d(z)}$	9.33±0.33$^{b(x)}$	311.19±0.12$^{d(x)}$
Ascorbic acid (3%)	0	21.75±0.13$^{cd(x)}$	6.25±1.23$^{b(z)}$	310.38±0.11$^{bc(y)}$
	5	17.91±0.26$^{b(y)}$	7.19±1.02$^{c(y)}$	312.31±1.48$^{c(x)}$
	10	11.25±0.22$^{c(z)}$	8.78±0.64$^{c(x)}$	313.23±0.19$^{c(x)}$
Citric acid (3%)	0	21.42±0.12$^{b(x)}$	6.69±0.72$^{b(z)}$	310.83±0.23$^{bc(z)}$
	5	17.55±1.17$^{b(y)}$	7.13±0.13$^{c(y)}$	312.34±1.75$^{c(y)}$
	10	11.19±0.85$^{c(z)}$	8.15±0.67$^{c(x)}$	315.25±0.56$^{c(x)}$
AA (1.5%)-CA (1.5%)	0	22.11±1.44$^{b(x)}$	6.21±0.12$^{b(y)}$	311.21±1.27$^{b(y)}$
	5	18.54±0.19$^{b(y)}$	8.19±0.44$^{b(x)}$	314.48±0.09$^{c(y)}$
	10	13.28±0.87$^{b(z)}$	8.48±0.15$^{c(x)}$	316.13±1.29$^{b(x)}$
Ultrasound (15 min)	0	28.19±1.14$^{a(x)}$	8.05±0.34$^{d(y)}$	315.68±1.12$^{a(z)}$
	5	24.11±0.13$^{a(y)}$	10.18±1.12$^{a(x)}$	319.12±0.68$^{a(y)}$
	10	19.24±1.29$^{a(z)}$	10.72±1.25$^{a(x)}$	325.65±0.74$^{a(x)}$

$^{a-e}$ Treatment means showed the effect of different treatments for the same day are not significantly different (p<0.05).

$^{x-z}$ Treatment means showed the effect of storage times for the same treatment are not significantly different (p<0.05).

Final Remarks

US was reported to be the most efficient application rather than chemical treatments to inhibit browning in fresh-cut radish slices. In addition to enzyme inactivation, US showed the best results on the inhibition of color degradation of fresh-cut radish samples. Furthermore, fresh-cut radish samples exposed to US treatment exhibited better AAC, ATC, and TPC. Fresh-cut radish pieces can be treated with US to prevent browning, rotting, and degradation. The current research is important in terms of the production of minimally processed radish with an enhanced product life and higher customer desirability by retaining the color and restricting the enzymatic browning.

References

[1] Huxsoll CC, Bolin HR. 1989. Processing and Distribution Alternatives for Minimally Processed Fruits and Vegetables. *Food Technol*, 43, 124-8.

[2] Marshall MR, Kim JM, Wei CI. 2000. *Enzymatic Browning in Fruits, Vegetables and Seafoods.* http://www.fao.org/ag/ags/agsi/ENZYMEFINAL/Enzymatic%20Browning.html

[3] Artes F, Castaner M, Gil MI. 1998. El pardeamiento enzim tico en frutas y hortalizas minimamente procesadas. *Food Sci Res Int*, 4, 377-389.

[4] Wiley RC. 1994. *Minimally processed refrigerated fruits and vegetables.* Chapman and Hall, New York.

[5] Lee SK, Kader AA. 2000. Preharvest and postharvest factors influencing vitamin C content of horticultural crops. *Postharvest Biol Technol*, 20, 207-220. https://doi.org/10.1016/S0925-5214(00)00133-2

[6] Queiroz C, Lopes MLM, Fialho E, Valente-Mesquita ML. 2008. Polyphenol oxidase: Characteristics and mechanisms of browning control. *Food Rev Int*, 24, 361-375. https://doi.org/10.1080/87559 12080 2089332

[7] Rico D, Mart n-Diana AB, Barat JM, Barry-Ryan C. 2007. Extending and measuring the quality of fresh-cut fruit and vegetables: A review. *Trends Food Sci Technol*, 18, 373-386. https://doi.org/10.1016/j.tifs.2007.03.011

[8] Yamaguchi K, Kato T, Noma S, Igura N, Shimoda M. 2010. The effects of high hydrostatic pressure treatment on the flavor and color of grated ginger. *Biosci Biotechnol Biochem*, 74, 1981-1986. https://doi.org/10.1271/bbb.90712

[9] Schilling S, Schmid S, Jaeger H, Ludwig M, Dietrich H, Toepfl S, Carlet R. 2008. Comparative study of pulsed electric field and thermal processing of apple juice with particular consideration of juice quality and enzyme deactivation. *J Agric Food Chem*, 56, 4545-4554. https://doi.org/10.1021/jf0732713

[10] Aadil RM, Wang MW, Liu ZW, Han Z, Zhang ZH, Jing H, Jabbar S. 2015. A potential of ultrasound on minerals, microorganism, phenolics and coloring pigments of grapefruit juice. *International J Food Sci Technol*, 50(5), 1144-1150. https://doi.org/10.1111/ijfs.12767

[11] Yildiz G, Izli G, Aadil RM. 2020. Comparison of chemical, physical, and ultrasound treatments on the shelf life of fresh-cut quince fruit (Cydonia oblonga Mill). *J Food Process Preserv*, 44 (3), e14366. https://doi.org/10.1111/jfpp.14366

[12] Minami K, Netto JT. 1994. Cultura do rabanete. ESALQ – Departamento de Horticultura, Piracicaba.

[13] Cruz-Cansino NS, Ramírez-Moreno E, León-Rivera J, Delgado-Olivares L, Alanís-García E, Ariza-Ortega JA, Manríquez-Torres JJ, Jaramillo-Bustos DP, 2015. Shelf life, physicochemical, microbiological and antioxidant properties of purple cactus pear (Opuntia ficus indica) juice after thermoultrasound treatment. *Ultrason Sonochem*, 27, 277-286. https://doi.org/10.1016/j.ultsonch.2015.05.040

[14] Rangana S. 1986. *Handbook of analysis and quality control for fruit and vegetable products*. 2nd ed. New Delhi, India: Tata Mc Graw Hill Publishing Co.

[15] Igual M, García-Martínez E, Martín-Esparza ME, Martínez-Navarrete N. 2012. Effect of processing on the drying kinetics and functional value of dried apricot. *Food Res Int*, 47, 284-290. https://doi.org/10.1016/j.foodres.2011.07.019

[16] Yildiz G, Aadil RM. 2022. Comparative Analysis of Antibrowning Agents, Hot Water and High-intensity Ultrasound Treatments to Maintain the Quality of Fresh-cut Mangoes. *J Food Sci Technol*, 59, 202-211. https://doi.org/10.1007/s13197-021-05001-y

[17] Wang Y, Zhang M, Mujumdar AS. 2011. Trends in processing technologies for dried aquatic products. *Dry Technol*, 29, 382-394. https://doi.org/10.1080/07373937.2011.551624

[18] Teng Y, Murtaza A, Iqbal A, Fu J, Ali SW, Iqbal MA, Xu X, Pan S, Hu W. 2020. Eugenol emulsions affect the browning processes, and microbial and chemical qualities of fresh-cut Chinese water chestnut. *Food Biosci*, 38, 100716. https://doi.org/10.1016/j.fbio.2020.100716

[19] Putnik P, Pavlić B, Šojić B, Zavadlav S, Žuntar I, Kao L, Kitonić D, Kovačević DB. 2020. Innovative Hurdle Technologies for the Preservation of Functional Fruit Juices. *Foods*, 9, 699. https://doi.org/10.3390/foods9060699

[20] Kasım R, Kasım MU. 2015. Biochemical changes and color properties of fresh-cut green bean (Phaseolus vulgaris L. cv.gina) treated with calcium chloride during storage. *Ciencia Tecnol Alime*, 35(2), 266-272. https://doi.org/10.1590/1678-457X.6523

[21] Prakash A, Baskaran R, Vadivel V. 2020. Citral nanoemulsion incorporated edible coating to extend the shelf life of fresh cut pineapples, *LWT-Food Sci Tehnol*, 118, 108851 https://doi.org/10.1016/j.lwt.2019.108851

[22] Erihemu, Wang M, Zhang F, Wang D, Zhao M, Cui N, Gao G, Guo J, Zhang Q. 2021. Optimization of the process parameters of ultrasound on inhibition of polyphenol oxidase activity in whole potato tuber by response surface methodology. *LWT – Food Sci Technol,* 144, 111232. https://doi.org/10.1016/j.lwt.2021.111232

[23] Fan K, Zhang M, Chen H. 2020. Effect of Ultrasound Treatment Combined with Carbon Dots Coating on the Microbial and Physicochemical Quality of Fresh-Cut Cucumber. *Food Bioprocess Technol,* 13, 648-660. https://doi.org/10.1007/s11947-020-02424-x

[24] Tapia MR, Gutierrez-Pacheco MM, Vazquez-Armenta FJ, Gonz.lez-Aguilar GA, Ayala Zavala JF, Rahman MSH, Siddiqui MW. 2015. Washing, peeling and cutting of fresh-cut fruits and vegetables, in: *Minimally processed foods.* MW Siddiqui, MS Rahman (Eds.), Springer, Switzerland, pp. 57-58.

[25] Yadav AK, Singh SV. 2014. Osmotic dehydration of fruits and vegetables: A review. *J Food Sci Technol,* 51, 1654-1673. https://doi.org/10.1007/s13197-012-0659-2

[26] Gallo M, Ferrara L, Naviglio D. 2018. Application of Ultrasound in Food Science and Technology: A Perspective. *Foods,* 7(10), 164. https://doi.org/10.3390/foods7100164

[27] Taha A, Ahmed E, Ismaiel A, Ashokkumar M, Xu X, Pan S, Hu H. 2020. Ultrasonic emulsification: An overview on the preparation of different emulsifiers-stabilized emulsions. *Trends Food Sci Technol,* 105, 363-377. https://doi.org/10.1016/j.tifs.2020.09.024

[28] Yao Y. 2016. Enhancement of mass transfer by ultrasound: Application to adsorbent regeneration and food drying/dehydration. *Ultrason Sonochem,* 31, 512-531. https://doi.org/10.1016/j.ultsonch.2016.01.039

[29] Daisy LL, Nduko JM, Joseph WM. 2020. Effect of edible gum Arabic coating on the shelf life and quality of mangoes (Mangifera indica) during storage. *J Food Sci Technol,* 57, 79-85. https://doi.org/10.1007/s13197-019-04032-w

[30] De Sousa AED, Fonseca KS, da Silva Gomes WK, da Silva APM, de Oliveira Silva E, Puschmann R. 2017. Control of browning of minimally processed mangoes subjected to ultraviolet radiation pulses. *J Food Sci Technol,* 54, 253-259. https://doi.org/10.1007/s13197-016-2457-8

[31] Jiang Q, Zhang M, Xu B. 2020. Application of ultrasonic technology in postharvested fruits and vegetables storage: A review. *Ultrason Sonochem,* 69, 105261. https://doi.org/10.1016/j.ultsonch.2020.105261

[32] Pawar SV, Rathod VK. 2020. Role of ultrasound in assisted fermentation technologies for process enhancements. *Prep Biochem Biotechnol,* 50(6), 627-634. https://doi.org/10.1080/10826068.2020.1725773

[33] Bhat R, Kamaruddin NS, Min-Tze L, Karim AA. 2011. Sonication improves kasturi lime (Citrus microcarpa) juice quality. *Ultrason Sonochem,* 18, 1295-1300. https://doi.org/10.1016/j.ultsonch.2011.04.002

[34] Lurie S. 2001. Physical treatments as replacements for postharvest chemical treatments. *Acta Hortic*, 553, 533-536. https://doi.org/10.17660/ActaHortic.2001.553.124

[35] Abid M, Jabbar S, Wu T, Hashim MM, Hu B, Saeeduddin M, Zeng X. 2015. Qualitative Assessment of Sonicated Apple Juice during Storage. *J Food Process Pres*, 39, 1299-1308. https://doi.org/10.1111/jfpp.12348

Chapter 7

Comparative Analysis of Ultrasonic Probe and Water Bath Systems on the Functional Properties of Whey Protein Isolate

Menekse Bulut[1]
Rana Muhammad Aadil[2]
and Gulcin Yildiz[1,*]

[1] Igdir University, Faculty of Engineering, Food Engineering Department, Iğdır, Turkey
[2] National Institute of Food Science and Technology, University of Agriculture, Faisalabad, Pakistan

Abstract

The current research was conducted for the investigation of the effects of high-intensity ultrasound (HIU) on the functional properties of the whey protein isolates (WPI). HIU was applied in two ways: ultrasonic bath and ultrasonic probe. In an ultrasonic water bath system, WPI was treated with 40 kHz frequency & power of 200 W with various time periods (10, 20, and 30 min) at 30°C, while ultrasound probe treatment was given through a VC-500 ultrasound generator at 20 kHz for 5 min at 30°C. The highest water solubility profile was obtained in WPI samples exposed to the ultrasound probe at 30 min (HIU4) (38.82%). A significant reduction in the size of WPI was achieved for HIU4 (241.12 nm). Furthermore, HIU4 samples resulted in the highest surface hydrophobicity and free sulfhydryl groups. The apparent viscosities of HIU4 samples (238.0 mPa.s) were found much higher compared to the other treatments. It was also observed that the HIU probe treatment has obtained better results than the ultrasonic water bath treatment.

[*] Corresponding Author's Email: gulcn86@gmail.com.

In: Power Ultrasound and Its Applications in Food Processing
Editors: Gulcin Yildiz and Gökçen Yıldız
ISBN: 979-8-88697-639-7
© 2023 Nova Science Publishers, Inc.

Keywords: high-intensity ultrasound, microstructure, protein solubility, surface hydrophobicity, whey protein isolate

Introduction

Whey protein is a crucial material of functional protein components for several traditional, and novel food materials [1]. Whey protein sources are recognized as complete proteins since they include 9 essential amino acids. The lactose content is low in whey products. When the liquid whey is obtained as a by-product of cheese or yoghurt fabrication, it is subjected to different processes in order to make the protein content higher [2]. After enough protein concentration is obtained, the liquid could be dried to develop whey concentrate powder including nearly 80% protein. On the other hand, if different processing processes are utilized to decrease the fat and carbohydrate content of whey, a whey isolate powder that consists of 90% or higher protein could be formed [2]. On the whole, steps that take place in the production of whey isolate lead to having more protein & less fat and carbohydrate content. Whey protein isolates (WPI) are a significant food ingredient due to their beneficial functional characteristics, including foaming, emulsification and gelation. İmportant proteins found in whey can be listed as bovine serum albumin (BSA), a-lactalbumin, b-lactoglobulin, & proteins are composed of almost 70 percent of all whey proteins. Those proteins are in charge of the functional features of WPI i.e., foaming, emulsification, texture, viscosity, gelation, color, water-solubility, flavor and improvement and propose various nutritional benefits to functionalized products [3].

The functional characteristics of WPI could be enhanced via a different method, for instance; the arrangement of covalent compounds along with reducing sugars and/or electrostatic aggregates with the polysaccharide [4], heat-induced polymerization [5], enzymatic hydrolysis [6], and other recent non-thermal methods including ultrasound, high-hydrostatical pressure, gamma irradiation, pulsed electric field, etc. [7]. HIU is a cost-effective and quick process that modify the functional and structural characteristics of the proteins [8]. In a study, ultrasound probe treatment attained significant results in apparent viscosity of WPI solutions [9]. Ultrasound probe pretreatment enhanced the solubility, foaming property and emulsifying attributes of several protein types including WPI [10]. Liu et al. [11] analyzed the impact of the water bath and ultrasound water bath. They observed that the antioxidant capacity of WPI was improved via ultrasound water bath as compared to the water bath. The impact of HIU is obtained via the physiochemical and mechanical impacts of acoustic cavitation. The cavitation is defined as the creation, development, and collapse of tiny-scale bubbles in the solutions. This phenomenon could be the reason for protein structure modification thanks to H_2 bonds and hydrophobic relationships and breakage of protein ingredients [12]. HIU was utilized to alter the functional, physical, and chemical features of native whey protein samples. So, the information related to HIU affects the functional characteristics of WPI and comparison of different HIU system is still limited.

If functional features of whey proteins are enhanced, whey proteins will provide desirable functionality to many food products, such as improved appearance, body, texture, and consistency. By taking care of the positive sides including cost-effectiveness, non-toxicity, quick and efficient application, it is anticipated to accomplish enhanced WPI functionality with the help of the HIU process. The purpose of the current research is to analyze the impacts of HIU on the functional attributes of WPI with ultrasonic bath & probe systems. Differences in functional characteristics of WPI were determined for solubility, turbidity, free sulfhydryl groups, particle size, surface hydrophobicity, interfacial properties and microstructure attributes. By achieving this, we attempt to discover a better method for enhancement of whey protein's characteristics to be used in the food industry with a suitable technique (bath or probe system) by HIU technology.

Materials and Methods

WPI Sample Preparation and HIU Treatment

WPI samples supplied from Dynamize (Dynamize Nutrition, ISO 100 Hydrolyzed). WPI consists of 82% protein on a dry base. All chemical substances were bought from Sigma-Aldrich (St. Louis, MO, USA), and Fisher Scientific (Pittsburgh, PA, USA). In the probe system, HIU treatment was arranged by a VC-500 power of 500W under a frequency of 20 kHz (Sonic & Material, Inc., USA) and a hundred percent amplitude for 5 min. The WPI (3 g) samples were blended with 100 mL distilled H_2O and stirred for 30 min under an ambient condition with a magnetic stirrer. The probe was located in the beaker and the beaker was located in the ice bath all along sonication to prevent temperature increase. For the ultrasonic water bath treatment, WPI samples were put in a 250 mL Erlenmeyer flask and immersed in a water bath (Wise clean, WUC-A10H) conducting at a frequency of 40 kHz & power of 200 W with several times consist of 10, 20, and 30 minutes. The experiments were performed at ambient temperature (30°C) (Figure 1). For evaluation of HIU effect, the same experimental steps were conducted with no HIU and called untreated WPI.

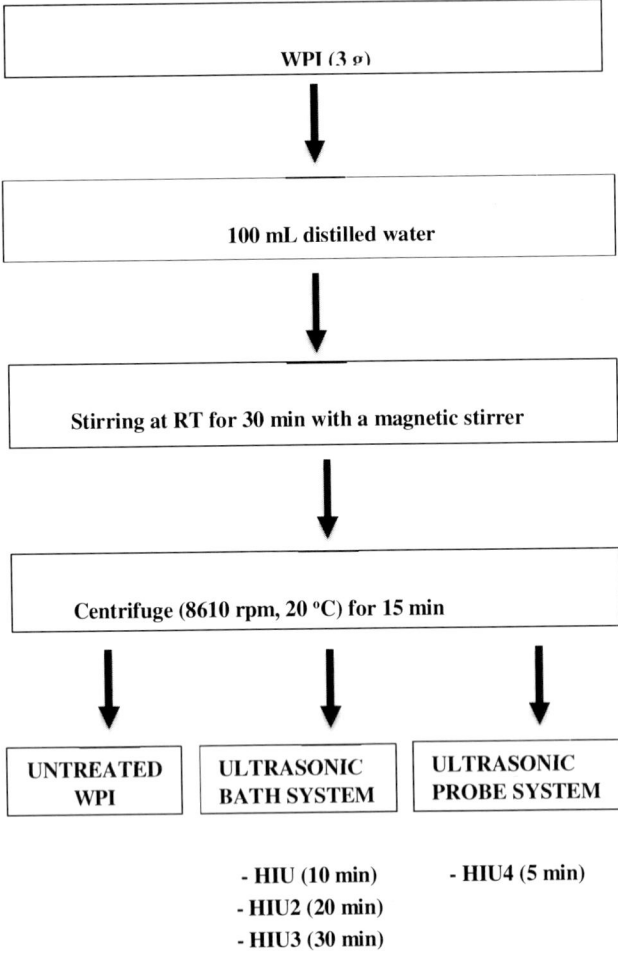

Figure 1. Preparation of WPI samples.

Protein Solubility

The protein concentration of soluble WPI aggregations was obtained by spectrophotometer using BSA as the standard [8]. Recovery of soluble protein was determined as below:

$$\text{Recovery of soluble protein} = \frac{\text{Protein concentration in soluble WPI}}{\text{Initial protein concentration}} \times 100$$

Particle Size and Turbidity

The volume-weighted particle sizes (D [4, 3]) of soluble WPI by dynamic light scattering (DLS) with the help of a NICOP 380 DLS instrument (Sant Barbara, CA, USA). Obtained WPI specimens subsequent to HIU treatment was diluted approximately 500-fold by using deionized water right before DLS analysis. Turbidity values of untreated and HIU-treated WPI solutions were analyzed with the assist of spectrophotometer (Lamba VIS/NIR Spectropmeter, Perkin Elmere, Watham, MA, USA) based on the approach proposed by Lee et al. [8]. DI water was utilized as a blank, and the absorbances were measured with a wavelength of 600 nm.

Surface Hydrophobicity (H_o)

H_o values of soluble whey protein aggregates were determined as stated by Yildiz et al. [12] with a slight change. The fluorescent probe used 1-anilin-8-naphthalene sulfonate (ANS) (Sigma Aldrich, St. Lois, MO, USA). ANS stock solution (8 mM) was arranged in phosphate solution (0.01 M; pH 7.0). Different whey protein isolate ratios (5 in total) changing from 0.04 to 0.2 mg/mL were also arranged with phosphate buffer solution (last volume 4 mL). A 20 µL ANS stock solution was blended with whey protein emulsions and fluorescence intensities were determined (Snyergy™2, BioTek Instrument Inc., Winooski, VT, USA) at 340 nm (excitation) and 440 nm (emission). Initial slopes of fluorescence intensity versus protein ratio were described as H_0 of whey protein isolates.

Free Sulfhydryl Group (Free SH) Measurement

The content of free SH was measured by a cysteine standard and these were arranged by following dissolved cysteine hydrochloride monohydrate at several concentrations including 1.5, 1.25, 1.0, 0.75, 0.5, 0.25 and 0.0 Mm in the reaction buffer solution. 50 mL Elman's reagent and 2.5 mL buffer solutions were arranged. Then, 250 mL of each standards were added to the test tube. After that, the mix in the test tube was blended and then incubated for 15 min at RT. Absorbance was measured at 412 nm. The values collected for the standards were plotted to develop a standard curve, and the specimen ratio was figured out from that curve. For the experiment in order to determine total SH content, 4 mL of Tris-Gly buffer was mixed with 1 mL of whey protein sources. The blend was incubated for around 60 min at 25°C. Following this, the blend was centrifuged (5000 g, 10 minutes). 0.04 mL of Elman's reagent solution was mixed with 4 mL of this solution, and the absorbances were determined at the wavelength of 412 nm. The half of values right after subtracting the SH values from the total SH values were described as the disulfide bonds (SS):

$$SS\ Content = \frac{Total\ SH - Free\ SH}{2} \left(\frac{\mu mol}{g}\right)$$

Interfacial Properties

Interfacial properties of WPI samples were determined via TAARES-G 2 Rheometer (TA Instrument, New Castle, DE, USA) following the technique stated by Yildiz et al. [12]. Protein solutions of WPI were made in deionized H_2O. All the interfacial variables were achieved at ambient temperature. Concentric cylinder geometry was utilized in order to measure the rheological properties for both contol and WPIs exposed to HIU process. The flow curve analysis was obtained by using a steady state flow ramp along with the shear rates changing from 1 to 100 s^{-1}. The shear rates were reported step by step with consecutive 180 seconds steps at constant shear rate. A 3 min was spent, because this time period is needed to get stable shear rate for each and every point. The viscosities of WPI samples were figured out for each point and recorded to create the flow curve graph.

Scanning Electron Microscope (SEM)

The morphological structure of the WPI specimens was investigated by using Scanning Electron Microscope (SEM). The WPI specimens were analyzed a wet mode. The WPI samples were frozen by using nitrogen right before SEM analysis. A small quantity of frozen WPI specimen was put into an aluminum stub and located in a vacuum chamber. The WPI specimen was explored by SEM microscope (SEM, Philips XL30 SEM-FEI, FGE Co., USA) along with a volt of 5.0 kV.

Statistical Analysis

The difference was determined with the General Linear Model process in SAS (version 9.3, SAS Institute, Incorporation, North Carolina, USA). A significant difference between the mean values was defined by Fisher's least significant difference test (LSD) of <0.05.

Results and Discussion

Solubility

Table 1 presents the findings related to water solubility values of the WPI samples employed by different HIU treatments. The highest soluble protein aggregates were detected for the HIU4 treatment (38.82%), whereas the lowest soluble proteins (3.89%) were found in the untreated WPI. HIU treatments, no matter if is it is a probe or bath system, had a significant improvement ($p < 0.05$) on the water solubility, but especially for HIU4 treatment. HIU-treated WPI with an ultrasonic bath at 10, 20, and 30 minutes exhibited significantly higher water solubility profile in comparison with the untreated WPI samples. On the other hand, those three showed significantly lower soluble protein profile with the comparison of HIU4 samples. While the WPI solubility was found between 9.41- 25.45% for ultrasonic bath treatments (HIU, HIU2, and HIU3), the WPI solubility was observed as 38.82% for the probe system (HIU4) (Table 1). Longer time leads to higher solubility for the ultrasonic bath system. While the solubility was found as 25.45% for 30 min treatment, it was significantly lower for 10- and 20-minutes treatments (9.41, and 19.24; subsequently). Solubility is a main functional property for a whey protein [13].

The solubility is about the several functional features such as molecular weight, not the primary but the secondary and tertiary structure, hydrophobicity, and electrostatic charges [8]. Processing treatments used to manufacture whey protein may result in heat-induced protein denaturation, which then reduces whey protein solubility. Native whey proteins remain soluble at around pH 7; however, heat-induced denaturation renders whey proteins less soluble than native whey proteins [14]. Thus, the protein solubility of whey protein is helpful to estimate protein denaturation [15]. Different works figured out that the enhancement of water solubility after an HIU process [8, 16]. Yildiz et al. [12] stated that the water solubilities of SPIs were significantly improved with the ultrasound treatment, from 9.08% to 82.5%. In a similar way, it was observed in the work of Lee et al. [8] where significantly higher soluble soy protein content was achieved with HIU application. The physical factors formed by ultrasonic cavitation effect such as shear forces could alter structural properties of protein components which comes out with enhanced protein solubility. Also, sonication can be the reason for the breakage of both non-covalent and covalent bonds which lead to improved soluble protein content [17]. Besides, Jiang et al. [16] pointed out that treated pea protein isolates (PPI) with an HIU showed a solubility seven times higher than the untreated PPI. Jambrak et al. [18] examined the influences of HIU (20 kHz probe & 40 kHz bath), on foaming attributes, emulsification, and water solubility of different whey protein types consist of a whey protein hydrolysate, whey protein concentrates and WPI. It was figured out whey protein solubility was significantly increased for all whey samples after both treatments (Probe: 20 kHz and water bath: 40 kHz) except for whey concentrates.

Table 1. Protein solubility, surface hydrophobicity, free-SH groups and particle size and turbidity of WPI samples

Treatment	Solubility (%)	Surface hydrophobicity (H_0)	Free SH (μmol g^{-1})	Particle size (nm)	Turbidity
Untreated	3.89±0.07[e]	194 ± 0.36[e]	2.47±0.71[e]	756.13 ± 0.38[a]	0.88 ±0.12[a]
HIU1	9.41±0.21[d]	215 ± 0.74[d]	2.66±0.18[d]	584.07 ± 0.69[b]	0.77 ±0.09[b]
HIU2	19.24±0.54[c]	297 ± 0.18[c]	3.12±0.22[c]	518.12 ± 0.74[c]	0.64 ±0.07[c]
HIU3	25.45±0.09[b]	350 ± 0.66[b]	3.98±0.09[b]	488.02 ± 0.26[d]	0.61 ±0.02[c]
HIU4	38.82±0.15[a]	423 ± 0.19[a]	5.12±0.73[a]	241.12 ± 0.58[e]	0.43 ±0.01[d]

[a-e]: Mean ± standard deviation (n=3) of properties with the same letter are not significantly different (p < 0.05).

*All the statistics were done separately for each parameter (solubility, surface hydrophobicity, and Free-SH).

Surface Hydrophobicity (Ho)

Ho value of WPI samples is shown in Table 1. There is a significant increase ($p < 0.05$) in all the ultrasonic applications in comparison with the untreated WPI sample. The lowest Ho (194.0) was found in the untreated WPI sample, while the highest Ho (423.0) was found after probe treatment (HIU4) in the WPI sample, which could be because of the collapse of gas bubbles which might create powerful shear forces and turbulence occurring with the ultrasonication. HIU-treatment broke the non-covalent bonds of WPI and make liable some of the hydrophobic parts that were formerly buried inside the WPI. Early works figured out HIU led to the exposure of some hydrophobic parts of proteins to the surface and created an increase in the H_o of various protein sources includes bovine serum albumin [19], soy protein isolate [20, 21]. A positive relationship was found between water solubility and H_0 (Table 1). For instance, HIU4 samples resulted with the highest water solubility (38.82%), which also showed the highest hydrophobicity (423.0). Similarly, the untreated WPI samples exhibited the lowest water solubility (3.89%), and the surface hydrophobicity values (194.0) were the lowest as well. This discovery is approved by an investigation of Yildiz et al. [12] who observed a positive interaction between the water solubility and surface hydrophobicity of SPIs. It was proved one more time in the studies of Lee et al. [8] and Jian et al. [16]. Both surface hydrophobicity and water solubility variables are significant factors affecting the emulsification stabilities and activities of protein sources [16]. Better emulsification and foam properties are outcomes of a balance that observed between hydrophilic and hydrophobic bonds [18]. WPI samples subjected to HIU process (especially, HIU4) showed both more surface hydrophobicity as well as high solubility, which could be a sign of enhanced emulsifying activity and protein stability.

Free Sulfhydryl Groups

Free SH content of control and HIU-treated WPIs is shown in Table 1. There is a significant increase ($p < 0.05$) in all ultrasonic treatments in comparison with the untreated WPI sample while ultrasonic probe treatment (5.12 µmol/g) have the highest values than all other ultrasound and untreated values (2.47 µmol/g). A higher SH group presents mostly subjected to internal SH group due to the unfolded protein response led via ultrasonic cavitation effect. Therefore, the surface SH groups rely on the conformation changes and protein unfolding (Jian et al., 2017). The rising in the free SH group might also be built by smaller WPI particles following ultrasound application, which leads to the buried SH group in WPI to be exposed to the external face [16]. An increment in the content of free SH group is additionally supported by Lee et al. [8] in the HIU-treated soy protein isolates with the comparison of the control treatment. In the current study, HIU4-treatment had the highest soluble protein (38.82%) which demonstrates the increase in SH groups that leads to an increase in the water solubility of the WPI sample.

Particle Size and Turbidity

DLS results of WPIs are exhibited in Table 1. The largest particle sizes were detected for untreated WPI samples (756.13 nm) while the smallest particle size was observed in the ultrasonic probe method (HIU4). WPI samples treated in an ultrasonic bath (HIU, HIU2, and HIU3) also exhibited the smaller particle size in comparison with the WPI samples without HIU application. But their sizes were significantly larger than the HIU4 samples. Similarly, in the study of Jambrak et al. [14] following application with an ultrasonic probe (20 kHz), HIU led to a decrease in particle size as well as narrowed the distribution, and significantly raise specific free surface ($p < 0.05$) in all WPI specimens. Following an application by an ultrasonic water bath (40 kHz), there was a significant ($p < 0.05$) decline in the particle sizes, but not to the extent of the ultrasonic probe method. The bath is mostly employed to clean the food products; its energy is transmitted into a liquid dispersion from a piezoelectric transducer placed at the bottom of the tank, which assures that energy is more dispersed. The probe, on the other hand, is usually utilized in order to break the cell membranes down, thus freeing the intracellular ingredient to be explored. The probe is fixed to the end of the transducer amplifier and has a direct connection with the sample. So, it is recognized as more effective compared to the bath, as the energy of the bath is in a more disperse structure [22]. The declines in the particle sizes of plant-based protein sources were observed after ultrasound treatment in various researches [14-16]. When the user in protein suspensions, ultrasound treatment was expressed to significantly lower the particle size of WPI [23]. Moreover, Karki et al. [24] determined that the particle sizes of defatted soy flake samples were decreased approximately 10-fold after HIU application. It was figured out that the cavitation may be the explanation of the breakage of protein substances, and decline in the particle size. Gordan and Pilosopf [25] accomplished to control particle size via HIU by merging several treatment periods, temperatures and ratios of WPI dispersions. Higher the treatment time, lower the particle size and particle size distribution. HIU develops a new surface and makes lower the size of the aggregates. By lowering the particle sizes, the free surface of the sample increases. In the case of this scenario, the protein particles are decreased due to the cavitation phenomena. This involves the degradation of protein aggregates and agglomerates. Ultrasonic cavitation is very efficient to break up protein sources and smaller particle aggregates the van der Walls's forces [14].

Figure 2. The appearance of untreated and HIU3 and HIU4 treated WPI samples.

The turbidity findings of control and HIU-treated WPIs are demonstrated in Table 1 and the appearance of control, HIU3 & HIU4-treated WPI samples is exhibited in Figure 2. Both the solubility and particle size of soluble protein aggregates determines the turbidity of protein dispersions [8]. Martini et al. [26] handled powerful ultrasonic sound waves to lessen the turbidity of whey solutions. It was concluded that around 90% decrease was observed in the turbidity of samples treated with ultrasound processing. The highest declines in turbidity values were determined when HIU was conducted for 15 min by using 15W of electrical power under 60°C. As HIU4 treatment shown the lowest turbidity (0.43), while the highest turbidity (0.88) with the lowest solubility (3.89%) was observed for the untreated WPI (Table 1). The HIU4 sample had a high protein concentration, but since HIU4 sample also demonstrated the smallest particle size, its turbidity value was resulted with the lowest (0.43). The HIU4 samples exhibited an almost transparent appearance (Figure 2). On the contrary, the untreated WPI samples seem cloudy as they show the largest particle sizes which are around 756.13 nm (Table 1). Sonication at 20 kHz increased the clarity of whey protein suspensions mostly because of the decrease in the sizes of the suspended insoluble protein ingredients.

Interfacial Properties

The flow curve (apparent viscosity vs. shear ratio) of control and WPI specimens subjected to HIU process are demonstrated in Figure 3. The data was obtained at room temperature at 6 different shear rates (0, 20, 40, 60, 80 and 100/s). The power-law coefficients (n_{app}, n, k and r^2) are tabulated in Table 2. By taking into account of Figure 3, the apparent viscosity of the WPI samples exposed to the HIU process decreased when the shear rate increases, indicating the protein dispersions have a shear-thinning or pseudo-plastic behavior. This is in agreement with their flow behavior variables. At Table 2, both untreated and treated WPI dispersions with HIU process have "n" less than 1, showing the pseudoplastic behavior. The apparent viscosities of HIU4 samples (238.0 mPa.s) was found much higher compared to the other treatments for all shear rates. This could be related to the improved solubility and hence high number densities of HIU4 samples. Shear thinning behavior is generally resulted from the breakage of protein sources, alignment of the aggregates in the flow direction, or disruption of both covalent bonds and noncovalent interactive forces consist of H_2 bonds, hydrophobic, and electrostatic interactions. And this might lead to change on protein network structure. The shear-thinning tendency of WPI dispersions resulted from HIU effect could be based on the powerful cavitation phenomena facilitating the mentioned modification in the WPI samples.

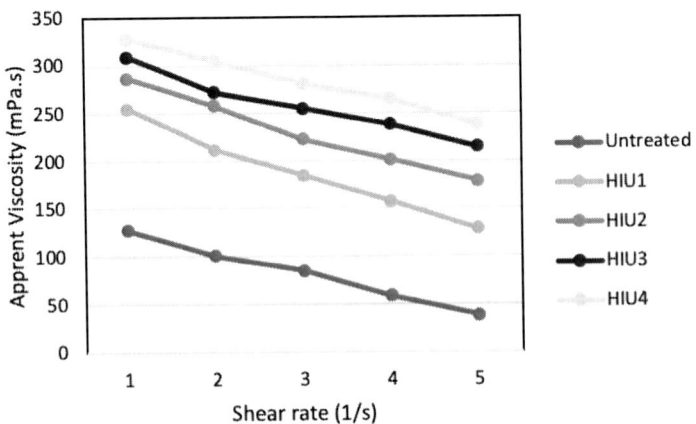

Figure 3. The flow curves (apparent viscosity versus shear rate) of the untreated and HIU-treated WPIs.

Table 2. Interfacial properties (apparent viscosity (n_{app}), flow behavior indices (n), consistency coefficients (k) and the regression coefficients (r^2) of WPI samples

Treatment	Apparent viscosity (n_{app}) mPas	Consistency coefficients (k) mPas	Flow behavior indices (n)	Regression coefficients (r^2)
Untreated	38.0 ± 0.7^e	44.2 ± 0.6^e	0.864 ± 0.01^a	0.997
HIU1	129.0 ± 0.1^d	$21 \times 10^3 \pm 0.7^d$	0.366 ± 0.08^b	0.995
HIU2	179.0 ± 0.4^c	$27 \times 10^3 \pm 2.8^c$	0.312 ± 0.22^b	0.995
HIU3	215.0 ± 0.9^b	$4.35 \times 10^3 \pm 2.6^b$	0.298 ± 0.19^b	0.994
HIU4	238.0 ± 0.5^a	$2.076 \times 10^3 \pm 1.9^a$	0.212 ± 0.23^b	0.991

$^{a-e}$: Mean ± standard deviation (n = 3) of properties with the same letter are not significantly different (p < 0.05).

* All the statistics were done separately for each parameter (apparent viscosity (n_{app}), flow behavior indices (n), consistency coefficients (k) and the regression coefficients (r^2).

Scanning Electron Microscope (SEM)

SEM images (1 µm) of the untreated & HIU-treated WPI samples are shown in Figure 4. Both untreated and HIU-treated WPI samples presented a spherical microstructure with particle sizes closely corresponding to the findings collected by dynamic light scattering (Table 1). Untreated WPI samples showed larger particles in comparison with the HIU-treated WPI samples. Following the application using a probe system with a 20 kHz, HIU lead to a decrease in particle size (p < 0.05) in WPI samples. On the other hand, at the end of application with water bath system during 30 min, there was a significant (p < 0.05) decrease too in the droplet size of WPI samples, however not to the extent as probe method.

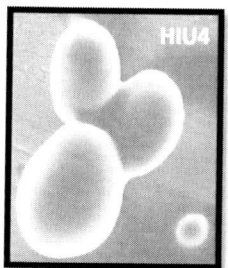 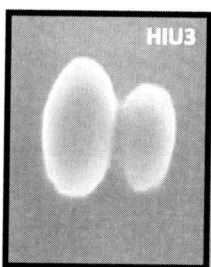 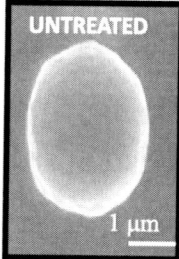

Figure 4. SEM images of untreated and HIU-treated WPI with an ultrasonic bath and ultrasonic probe systems.

Conclusion

Both ultrasonic bath and probe treatments were effective on the modification of whey protein properties. The study shows that the ultrasound application with a probe system significantly increased solubility of commercial WPI, reduced the sizes of protein aggregates to less than 300 nm, and resulted with enhanced functional properties. This study demonstrated that the HIU4 application might be a useful means to modify whey protein isolates for the enhancement of the functional properties. The functionalized WPI produced by HIU4 treatment can be used in liquid food with high solubility and less precipitation. Overall, WPI samples treated by HIU4 treatment could be utilized as a wall material to encapsulate the bioactive metabolites for the management of functional food products with improved features. This expresses that the soluble whey protein aggregates produced by HIU4 might be a promising alternative for the preparation of nanoparticles and nanostructures in order to carry and provide the retention of secondary metabolites.

References

[1] Kumar, R., Chauhan, S.K., Shinde, G., Subramanian, V., & Nadanasabapathi, S. (2018). Whey Proteins: A potential ingredient for food industry- A review, *Asian Journal of Dairy and Food Research*, 37, 283-290. http://dx.doi.org/10.18805/ajdfr.DR-1389

[2] Liu, Q., Li, J., Kong, B. H., Li, P. J., & Xia, X. F. (2014). Physicochemical and antioxidant properties of Maillard reaction products formed by heating whey protein isolate and reducing sugars. *International Journal of Dairy Technology*, 67, 220-228. https://doi.org/10.1111/1471-0307.12110

[3] KrešicÅL, G., Lelas, V., Jambrak, A.R., Herceg, Z., & BrncˇicÅL, S.R. (2008). Influence of novel food processing technologies on the rheological and thermophysical properties of whey proteins. *Journal of Food Engineering*, 87 (1), 64–73. http://dx.doi.org/10.1016%2Fj.jfoodeng.2007.10.024

[4] Dickinson, E. (2015). Colloids in food: Ingredients, structure, and stability. *Annual Review of Food Science and Technology*, 6, 211-233. https://doi.org/10.1146/annurev-food-022814-015651

[5] Ryan, K. N., & Foegeding, E. A. (2015). Formation of soluble whey protein aggregates and their stability in beverages. *Food Hydrocolloids*, 43, 265-274. https://doi.org/10.1016/j.foodhyd.2014.05.025

[6] Wang, W., Zhong, Q., & Hu, Z. (2013). Nanoscale understanding of thermal aggregation of whey protein pretreated by transglutaminase. *Journal of Agricultural and Food Chemistry*, 61, 435-446. https://doi.org/10.1021/jf304506n

[7] Mirmoghtadaie, L., Aliabadi, S. S., & Hosseini, S. M. (2016). Recent approaches in physical modification of protein functionality. *Food Chemistry*, 199, 619-627. https://doi.org/10.1016/j.foodchem.2015.12.067

[8] Lee, H., Yildiz, G., Dos Santos, L.C., Jiang, S., Andrade, J., Engeseth, N.C., & Feng, H. (2016). Soy protein nano-aggregates with improved functional properties prepared by sequential pH treatment and ultrasonication. *Food Hydrocolloids*, 55, 200–209. https://doi.org/10.1016/j.foodhyd.2015.11.022

[9] Shen, X., Shao, S., & Guo, M. (2017). Ultrasound-induced changes in physical and functional properties of whey proteins. *International Journal of Food Science & Technology*, 52(2), 381-388. https://doi.org/10.1111/ijfs.13292

[10] Sun, Y., Chen, J., Zhang, S., Li, H., Lu, J., Liu, L., Uluko, H., Su, Y., Cui, W., Ge, W., & Lv, J. (2014). Effect of power ultrasound pre-treatment on the physical and functional properties of reconstituted milk protein concentrate. *Journal of Food Engineering*, 124, 11-18. https://doi.org/10.1016/j.jfoodeng.2013.09.013

[11] Liu, L., Li, X., Du, L., Zhang, X., Yang, W., & Zhang, H. (2019). Effect of ultrasound assisted heating on structure and antioxidant activity of whey protein peptide grafted with galactose. *LWT-Food Science and Technology*, 109, 130-136. https://doi.org/10.1016/j.lwt.2019.04.015

[12] Yildiz, G., Andrade, J., Engeseth, N.C., & Feng, H. (2017). Functionalizing soy protein nano-aggregates with pH-shifting and mano-thermo-sonication. *Journal of Colloid and Interface Science*, 505, 836-846. https://doi.org/10.1016/j.jcis.2017.06.088

[13] Hussain, R., Gaiani, C., Jeandel, C., Ghanbaja, J., & Scher, J. (2012). Combined effect of heat treatment and ionic strength on the functionality of whey proteins. *Journal of Dairy Science*, 95, 6260-6273. https://doi.org/10.3168/jds.2012-5416

[14] Jambrak, A.R., Mason, T.J., Lelas, V., Paniwnyk, L., & Herceg, Z. (2014). Effect of ultrasound treatment on particle size and molecular weight of whey proteins. *Journal of Food Engineering*, 121, 15-23. https://doi.org/10.1016/j.jfoodeng.2013.08.012

[15] Morr, C.V., & Ha, E.Y.W. (1993). Whey protein concentrates and isolates: processing and functional properties. *Critical Reviews in Food Science and Nutrition*, 33, 431–476. https://doi.org/10.1080/10408399309527643

[16] Jiang S, Ding J, Andrade J, Rababah T.M., Almajval, A., Abulmeaty, M.M., & Feng, H. (2017). Modifying the physicochemical properties of pea protein by pH-shifting and ultrasound combined treatments. *Ultrasonics Sonochemistry*. 38, 835-842. https://doi.org/10.1016/j.ultsonch.2017.03.046

[17] Hu, H., Li-Chen, E.C.Y., Wan, L., Tian, M., & Pan, S. (2003). The effect of high intensity ultrasonic pretreatment on the properties of soybean protein isolate gel induced by calcium sulfate. *Food Hydrocolloids*, 32 (2), 303–311. https://doi.org/10.1016/j.foodhyd.2013.01.016

[18] Jambrak, A.R., Mason, T.M., Lelas, V., Herceg, Z., & Herceg, I.L. (2008). Effect of ultrasound treatment on solubility and foaming properties of whey protein

suspensions. *Journal of Food Engineering*, 86, 281–287. https://doi.org/10.1016/j.jfoodeng.2007.10.004

[19] Gülseren, I., Güzey, D., Bruce, B.D., & Weiss, J. (2007). Structural and functional changes in ultrasonicated bovine serum albumin solutions. *Ultrasonics Sonochemistry*, 14 (2), 173–183. https://doi.org/10.1016/j.ultsonch.2005.07.006

[20] Chen, L., Chen, J., Ren, J., & Zhao, M. (2011). Effects of Ultrasound pretreatment on the enzymatic hydrolysis of soy protein isolates and on the emulsifying properties of hydrolysates. *Journal of Agricultural and Food Chemistry*, 59, 2600-2609. https://doi.org/10.1021/jf103771x

[21] Hu, H., Wu, J., Li-Chan, E. C. Y., Zhu, L., Zhang, F., Xu, X., Fan, G., Wang, L., Huang, X., & Pan, S. (2013). Effects of ultrasound on structural and physical properties of soy protein isolate (SPI) dispersions. *Food Hydrocolloids*, 30, 647-655. https://doi.org/10.1016/j.foodhyd.2012.08.001

[22] Venâncio, R. S. S., Tonello, P.S., Martins, A.C.G. (2015). Environmental technology: applications of ultrasound. *International Journal of Engineering and Applied Sciences*, 7 (2), 1-5.

[23] Jambrak, A. R., Lelas, V., Mason, T. J., Kresic, G., & Badanjak, M. (2009). Physical properties of ultrasound treated soy proteins. *Journal of Food Engineering*, 93, 386-393. https://doi.org/10.1016/j.jfoodeng.2009.02.001

[24] Karki, B., Lamsal, B. P., Jung, S., van Leeuwen, J. (Hans), Pometto, A. L., III, Grewell, D., & Khanal, S.K. (2010). Enhancing protein and sugar release from defatted soy flakes using ultrasound technology. *Journal of Food Engineering*, 96(2), 270-278. http://dx.doi.org/10.1016/j.jfoodeng.2009.07.023

[25] Gordon, L., & Pilosof, A. M. R. (2010). Application of high-intensity ultrasounds to control the size of whey proteins particles. *Food Biophysics*, 5, 203-210. https://doi.org/10.1007/S11483-010-9161-4

[26] Martini, S., Potter, R., & Walsh, M.K. (2010). Optimizing the use of power ultrasound to decrease turbidity in whey protein suspensions. *Food Research International*, 43, 2444–2451. https://doi.org/10.1016/j.foodres.2010.09.018

Index

A

antioxidant capacity (ATC), 42, 50, 103, 107, 108
antioxidants, 4, 11, 37, 38, 40, 42, 43, 50, 51, 52, 53, 57, 59, 60, 61, 64, 69, 70, 71, 73, 74, 99, 100, 101, 103, 107, 109, 114, 126, 127
apple-carrot juice, vii, ix, 37, 38, 41, 42, 43, 44, 45, 46, 47, 48, 49, 50, 51, 53, 54, 55, 56, 57, 58, 59

B

bioactive compounds, vii, ix, 10, 11, 12, 34, 35, 38, 39, 40, 60, 61, 62, 64, 66, 70, 72, 78, 90, 91, 92, 99, 100, 106, 107
bio-marine food, 79, 89, 90
bleaching, 1, 2, 4, 5, 7, 8, 9, 10, 14, 15
brining, 87
Brix values, 42
by-products, ix, 63, 64, 65, 67, 68, 69, 71, 72, 90, 91

C

carotenoids, 9, 28, 37, 38, 39, 44, 50, 51, 52, 54, 57, 60, 61, 64, 78, 91, 92
cavitation, 1, 3, 7, 10, 11, 17, 18, 21, 22, 23, 25, 26, 29, 32, 33, 39, 48, 50, 52, 53, 56, 64, 67, 68, 78, 83, 106, 107, 115, 120, 121, 122, 124
color values, 8, 9, 30, 46, 102, 103, 104, 106

D

deacidification, 1, 2, 4, 7, 10, 13
degumming, 1, 2, 4, 5, 6, 13

drying, vii, ix, 2, 11, 17, 18, 19, 26, 27, 28, 29, 30, 31, 32, 33, 34, 35, 36, 60, 72, 75, 85, 86, 95, 96, 109, 110, 133, 134

E

enzymes, 5, 21, 27, 39, 52, 56, 58, 62, 80, 83, 86, 94, 101

F

freeze drying, 26, 29
freezing, 21, 30, 36, 86

H

high power ultrasound (HPU), 81, 87, 88
high-intensity ultrasound, 33, 36, 113, 114, 128
hot air, 26, 27, 28, 30

I

infrared drying, 31, 32

L

lipids, 63, 78, 79, 91, 92
lower power ultrasound (LPU), 81

M

marine compounds, 78, 93
microbes, 58, 84, 85, 86, 94, 96
microstructure, 96, 114, 115, 125
microwave-assisted extraction (MAE), 66, 91, 92

O

osmotic drying, 26, 27

P

pasteurization, 18, 25, 37, 38, 40, 41, 57, 59, 84
pectin, vii, ix, 63, 64, 65, 66, 67, 68, 69, 70, 71, 72, 73, 74, 99
PME, 39, 102, 105, 106
PPO, 39, 52, 100, 101, 102, 105, 106
PPO activity, 101, 102, 105
protein solubility, 114, 117, 120
pulsed electric fields (PEF), 18, 84, 85, 88, 89, 90, 96, 97

Q

quality, vii, ix, 2, 3, 5, 6, 8, 10, 11, 12, 19, 26, 29, 32, 33, 34, 35, 37, 38, 39, 40, 41, 53, 57, 58, 59, 60, 61, 62, 75, 76, 77, 79, 80, 81, 86, 87, 89, 93, 94, 95, 96, 99, 100, 104, 108, 109, 110

R

radish, 99, 100, 101, 102, 103, 104, 105, 106, 107, 108

S

seafood processing, 75, 76, 77, 78, 79, 80, 90, 94
shelf-life, 38, 77, 99, 100, 104, 105
sonication, vii, ix, 10, 11, 37, 38, 39, 40, 41, 48, 49, 50, 52, 53, 55, 57, 59, 60, 68, 83, 84, 86, 101, 110, 115, 120, 123, 127
spoilage, 38, 48, 76, 78, 80, 86
spray drying, 29
surface hydrophobicity, 113, 114, 115, 117, 120, 121

T

thermosonication, 25, 37, 38, 39, 48, 58, 60, 61, 62, 81
total phenolic content (TPC), 42, 43, 50, 103, 107, 108
tropical fruits, 63, 64, 65, 66, 67, 69, 71, 100

U

ultrasonic bath, 3, 6, 8, 18, 24, 27, 28, 30, 81, 113, 115, 119, 122, 125, 126
ultrasonic probe, viii, ix, 3, 8, 18, 24, 30, 32, 66, 81, 113, 121, 122, 125
ultrasonic sound waves, 17, 18, 19, 20, 21, 22, 23, 24, 25, 26, 27, 28, 29, 30, 31, 32, 33, 123
ultrasound, iii, vii, ix, 1, 2, 3, 4, 6, 7, 8, 9, 10, 11, 12, 13, 14, 15, 17, 18, 19, 20, 21, 23, 25, 26, 28, 30, 31, 32, 33, 34, 35, 36, 39, 58, 59, 60, 61, 62, 63, 64, 66, 67, 68, 71, 72, 73, 75, 76, 78, 79, 80, 81, 82, 83, 86, 87, 88, 89, 94, 95, 96, 99, 100, 101, 103, 104, 105, 106, 107, 109, 110, 113, 114, 120, 121, 122, 123, 126, 127, 128, 133
ultrasound assisted freezing, 31, 76
ultrasound pre-treatment, vii, ix, 10, 11, 17, 18, 35, 101, 127
ultrasound treatment, vii, ix, 9, 11, 12, 59, 61, 62, 99, 100, 105, 106, 107, 109, 110, 120, 122, 127
ultrasound-assisted extraction, vii, ix, 63, 64, 66, 71, 73, 76, 88

V

vegetable oils, ix, 1, 2, 4, 13

W

whey protein isolate, viii, ix, 113, 114, 117, 126

About the Editors

Associate Professor Gulcin Yildiz, PhD
Igdir University, Faculty of Engineering, Food Engineering Department,
Iğdır, Turkey
Email: gulcn86@gmail.com

Dr. Gulcin Yildiz obtained her PhD in Food Science from University of Illinois at Urbana-Champaign, USA. She is currently working at food engineering department in Igdir University as an associate professor. She is also head of the department of gastronomy and culinary arts. She is the recipent of several prizes, awards and fellowships. She has attended several national/international scientific conferences and workshops as a speaker and shared her knowledge on food processing and engineering.

Interests: Novel food processing, power ultrasound, technological aspects (nutritional, microbial, enzymatic and chemical inactivation phenomena) in thermal and non-thermal processing studies, micro and nano-delivery systems, food drying, functional foods, fruit-vegetable technologies.

Associate Professor Gökçen Yıldız
Bursa Technical University, Faculty of Engineering and Natural Sciences,
Food Engineering Department, Bursa, Turkey
Email: gokcnyildiz@gmail.com

Gökçen Yıldız is an Associate Professor in the Department of Food Engineering at the Bursa Technical Univerity in Turkey. She has reviewed many manuscripts for publication in high-impact factor journals. She has several research and review articles (SCI Journals), book chapters and books in the field of food processing and technology. She teaches several undergraduate and postgraduate classes including fruit-vegetable technology, food chemistry, advanced sensory analysis, and drying technology. She has been involved in several projects as project manager and researcher.

Interests: Fruit-vegetable technology, drying process, food processing and preservation, functional foods, food chemistry.